TRANSPORTATION INFRASTRUCTURE:
ROADS, HIGHWAYS, BRIDGES, AIRPORTS AND MASS TRANSIT

MILEAGE FEES AND THE HIGHWAY TRUST FUND

AN EXAMINATION

TRANSPORTATION INFRASTRUCTURE: ROADS, HIGHWAYS, BRIDGES, AIRPORTS AND MASS TRANSIT

Additional books in this series can be found on Nova's website under the Series tab.

Additional E-books in this series can be found on Nova's website under the E-book tab.

U.S. TRANSIT, TRANSPORTATION AND INFRASTRUCTURE: CONSIDERATIONS AND DEVELOPMENTS

Additional books in this series can be found on Nova's website under the Series tab.

Additional E-books in this series can be found on Nova's website under the E-book tab.

TRANSPORTATION INFRASTRUCTURE:
ROADS, HIGHWAYS, BRIDGES, AIRPORTS AND MASS TRANSIT

MILEAGE FEES AND THE HIGHWAY TRUST FUND

AN EXAMINATION

YOAN BERGERON
EDITOR

nova publishers
New York

Library of Congress Cataloging-in-Publication Data

ISBN: 978-1-62618-829-7

Published by Nova Science Publishers, Inc. † New York

CONTENTS

PREFACE

Federal funding to build and maintain the nation's highways and bridges comes primarily from highway users through federal fuel taxes. These revenues have eroded due to improvements in vehicle fuel efficiency and other factors contributing to shortfalls in the Highway Trust Fund. Experts have proposed alternative means of raising revenues by charging drivers fees based on their miles traveled. Several states have tested systems that gather vehicle mileage and location data, which has raised privacy concerns. This book examines the benefits and challenges of mileage fee initiatives in the United States and other selected nations, and discusses whether mileage fee rates are necessary to replace and supplement current Highway Trust Fund revenues and the effect these fees would have on users' costs.

Chapter 1 – Federal funding to build and maintain the nation's highways and bridges comes primarily from highway users through federal fuel taxes. These revenues have eroded due to improvements in vehicle fuel efficiency and other factors contributing to shortfalls in the Highway Trust Fund. Experts have proposed alternative means of raising revenues by charging drivers fees based on their miles traveled. Several states have tested systems that gather vehicle mileage and location data, which has raised privacy concerns. GAO examined (1) the benefits and challenges of mileage fee initiatives in the United States and other selected nations, (2) mileage fee rates necessary to replace and supplement current Highway Trust Fund revenues and the effect these fees would have on users' costs, and (3) state DOTs' views on future revenue demands and mileage fees. GAO reviewed five domestic pilot projects and programs in Germany, New Zealand, and the Netherlands; modeled mileage fees for passenger vehicles and commercial trucks; and surveyed 51 state DOTs.

Chapter 2 – Federal funding for highways is provided to the states mostly through a series of grant programs known as the Federal-Aid Highway Program, administered by the Department of Transportation's (DOT) Federal Highway Administration (FHWA). In 2005, the Safe, Accountable, Flexible, Efficient Transportation Equity Act: A Legacy for Users (SAFETEA-LU) authorized $197.5 billion for the Federal-Aid Highway Program for fiscal years 2005 through 2009. The program operates on a "user pay" system, wherein users contribute to the Highway Trust Fund through fuel taxes and other fees. The distribution of funding among the states has been a contentious issue. States that receive less than highway users contribute are known as "donor" states and states that receive more than users contribute are known as "donee" states.

GAO was asked to examine for the SAFETEA-LU period (1) how contributions to the Highway Trust Fund compared with the funding states received, (2) what provisions were used to address rate-of-return issues across states, and (3) what additional factors affect the relationship between contributions to the Highway Trust Fund and the funding states receive. To conduct this review, GAO obtained and analyzed data from FHWA, reviewed FHWA and other reports, and interviewed FHWA and DOT officials.

GAO is not making any recommendations. DOT reviewed a draft of this report and provided technical comments, which we incorporated as appropriate.

In: Mileage Fees and the Highway Trust Fund ISBN: 978-1-62618-829-7
Editor: Yoan Bergeron

Chapter 1

HIGHWAY TRUST FUND: PILOT PROGRAM COULD HELP DETERMINE THE VIABILITY OF MILEAGE FEES FOR CERTAIN VEHICLES*

United States Government Accountability Office

WHY GAO DID THIS STUDY

Federal funding to build and maintain the nation's highways and bridges comes primarily from highway users through federal fuel taxes. These revenues have eroded due to improvements in vehicle fuel efficiency and other factors contributing to shortfalls in the Highway Trust Fund. Experts have proposed alternative means of raising revenues by charging drivers fees based on their miles traveled. Several states have tested systems that gather vehicle mileage and location data, which has raised privacy concerns. GAO examined (1) the benefits and challenges of mileage fee initiatives in the United States and other selected nations, (2) mileage fee rates necessary to replace and supplement current Highway Trust Fund revenues and the effect these fees would have on users' costs, and (3) state DOTs' views on future revenue demands and mileage fees. GAO reviewed five domestic pilot projects and programs in Germany, New Zealand, and the Netherlands; modeled mileage

* This is an edited, reformatted and augmented version of United States Government Accountability Office, Publication No. GAO-13-77, dated December 2012.

fees for passenger vehicles and commercial trucks; and surveyed 51 state DOTs.

What GAO Recommends

Should Congress further explore mileage fees, it should consider establishing a pilot program to test the viability of such fees for commercial trucks and electric vehicles. FHWA should update its estimates of road damages imposed by all vehicle types compared with the tax revenues generated by each. The Department of Transportation took no position on GAO's recommendation but provided technical comments which GAO incorporated as appropriate.

What GAO Found

Mileage-based user fee initiatives in the United States and abroad show that such fees can lead to more equitable and efficient use of roadways by charging drivers based on their actual road use and by providing pricing incentives to reduce road use. Mileage fees for passenger vehicles, however, continue to face significant public concerns related to privacy as well as cost challenges. Privacy concerns are particularly acute when Global Positioning System (GPS) units are used to track the location of passenger vehicles. Reliable cost estimates for mileage fee systems are not available, but implementing a system to collect fees from 230 million U.S. passenger vehicles is likely to greatly exceed the costs of collecting fuel taxes. Commercial truck user fee systems in Germany and New Zealand have achieved substantial revenues and benefits such as reduced road damage and emissions with fewer privacy concerns, but ensuring compliance in a cost effective manner presents trade-offs. Few commercial truck mileage fee pilots have been conducted in the United States, but efforts in two states suggest such fees pose fewer privacy and cost challenges than passenger vehicle fees.

Mileage fee rates could be set to replace or supplement current Highway Trust Fund revenues. GAO calculated average mileage fee rates for passenger vehicles and commercial trucks needed to meet three federal revenue targets ranging from $34 billion (replace current federal fuel tax revenues) to $78 billion (increase spending to maintain existing system conditions and performance). To meet these targets, drivers of passenger vehicles with average fuel efficiency would pay $108 to $248 per year in mileage fees compared to the $96 these drivers currently pay in federal gasoline tax. These fees would affect users' costs differently based on each vehicle's fuel efficiency, because drivers of less efficient vehicles now pay more in fuel taxes than drivers of vehicles with greater fuel efficiency. However, like federal fuel taxes, mileage fees would comprise a small portion of users' overall fuel costs and thus only marginally increase users' overall transportation costs. A mileage fee for commercial trucks could also increase users' costs, particularly for larger trucks that log more miles. In 2000, the Federal Highway Administration (FHWA) estimated that heavy commercial trucks generally pay less in federal taxes than the road damage costs they impose. Adjusting mileage fee rates to account for vehicle road damage costs would increase rates for commercial truck users. However, FHWA's estimates may not reflect current conditions. Setting rates to cover these costs would require updated estimates of vehicles' responsibility for road damage.

State departments of transportation (DOT) recognize the need for an alternative funding mechanism to meet future revenue demands, and many would support federal actions to evaluate mileage fees. Few states reported that they are likely to introduce such fees in the next 10 years, but more than half would support federally-led field tests of mileage fees for commercial trucks and electric vehicles. Although few electric vehicles are on the roads today, their numbers are expected to increase, and they do not contribute to the Highway Trust Fund. Without a federal pilot program to evaluate (1) options to more accurately charge commercial trucks and electric vehicles for their road use and (2) the costs and benefits of such systems, Congress lacks critical information to assess whether mileage fees for these vehicles could be a viable and cost-effective tool to help address the nation's surface transportation funding challenges.

ABBREVIATIONS

AASHTO	American Association of State Highway Transportation Officials
CAFE	Corporate Average Fuel Economy
CBO	Congressional Budget Office
DOT	department of transportation
E-RUC	Electronic Road User Charge

FHWA	Federal Highway Administration
GPS	Global Positioning System
HGV	Heavy Goods Vehicle
IFTA	International Fuel Tax Agreement
IRP	International Registration Plan
MAP-21	Moving Ahead for Progress in the 21st Century Act
MPG	miles per gallon
NCHRP	National Cooperative Highway Research Program
NSTIFC	National Surface Transportation Infrastructure Financing Commission
PSRC	Puget Sound Regional Council
RUC	Road User Charge
SAFETEA-LU	Safe, Accountable, Flexible, Efficient Transportation Equity Act: A Legacy for Users
SUV	sport utility vehicle
TRUE	Truck Road Use Electronics
VMT	vehicle miles traveled

December 13, 2012
The Honorable Tom Latham
Chairman

The Honorable John W. Olver
Ranking Member
Subcommittee on Transportation,
Housing and Urban Development, and Related Agencies
Committee on Appropriations
House of Representatives

The nation's surface transportation system is critical to the economy and affects the daily lives of most Americans. However, the system is under growing strain, and the costs of repairs and upgrades to meet current and future demands are estimated in the hundreds of billions of dollars. For more than 50 years, federal funding to build and maintain the nation's vast network of highways and bridges has been collected primarily from highway users through federal fuel taxes. This system is based on the "user pays" principle in which the costs of government programs or services, such as the construction and maintenance of roadways, are paid by the individuals and firms that use and benefit from the service through taxes or fees. For many years, user fees in the form of federal fuel taxes and taxes on commercial trucks provided sufficient revenues to the Highway Trust Fund, the primary source of federal funding for highway and transit programs.[1] However, revenues into the fund have eroded over time, in part because federal fuel tax rates have not increased since 1993 and in part because of improvements in vehicle fuel efficiency. This trend will continue in the years ahead as more alternative fuel vehicles take to the roads.

To maintain current spending levels and cover revenue shortfalls, Congress transferred more than $34 billion in general revenues to the Highway Trust Fund from fiscal year 2008 to 2010; in 2012, Congress appropriated an additional $18.8 billion in general revenues for fiscal years 2013 and 2014.[2] This approach has effectively broken the link between taxes paid and benefits received by users and may not be sustainable given competing demands and the federal government's growing fiscal challenge.[3] Although the President signed a 2-year surface transportation authorization in July 2012,[4] the Highway Trust Fund faces significant shortfalls in the years ahead, beginning in fiscal year 2015, to support current spending levels. Congress and the Administration have yet to develop a long-term plan for funding surface transportation; ultimately, increased surface transportation revenues, reduced transportation spending levels, or both will be needed to bring revenues and spending into balance. For this and other reasons, funding surface transportation remains on GAO's High-Risk List.[5]

To address this growing funding problem, transportation experts and economists have advocated developing an alternative revenue-generating system that preserves the user-pay principle by charging vehicle owners based on the number of miles they drive—that is, by their vehicle miles traveled. Such fees, commonly known as VMT fees or mileage-based user fees, are referred to as "mileage fees" in this report. Advocates of mileage fees argue that in addition to raising revenues, such fees may offer other benefits that could lead to more efficient use of the nation's highways and result in reduced congestion for drivers. For example, mileage fees and other forms of road pricing such as tolling send clear price signals to road users, and provide incentives to drivers to consider alternatives such as public transit or carpooling which can reduce congestion, vehicle emissions, and overall spending on fossil fuels. The Congressional Budget Office (CBO) reported that most drivers currently pay much less than the full cost of their highway use, and that mileage fees could provide a better incentive for efficient highway use than fuel taxes do because the majority of highway costs are related to miles driven.[6] In addition, we have reported that if those who benefit from a program do not bear the full social cost of the service, they may seek to have the government provide more of the service than is economically efficient.[7] Furthermore, two national transportation commissions established by Congress have reported that reforming highway financing with a mileage fee system could provide a more viable, long-term source of federal revenues than the current system of fuel and excise taxes.[8]

In recent years, several states have conducted pilot projects to test mileage fee systems and other countries have also evaluated or implemented mileage fees for light passenger vehicles or heavy commercial trucks. Efforts to evaluate mileage fees in the United States have been met with concerns from the general public, including the fear that installing Global Positioning System (GPS) technology into private vehicles to gather mileage data could also potentially be used to compromise drivers' personal information, such as locations visited.

In light of these concerns, you asked us to review the issues surrounding the possible use of a mileage-based user fee system to fund federal surface transportation programs. In this report we examine:

1. the benefits achieved and challenges faced in mileage fee initiatives in the United States and selected other nations,
2. the user fee rates necessary to replace and supplement current revenues to the Highway Trust Fund and the effect these fees would have on users' costs, and
3. the perspectives of state departments of transportation (DOT) on addressing future revenue demands using mileage fees.

To examine these issues, we gathered information on the benefits and challenges of several mileage fee pilot projects conducted in the United States, including efforts in Oregon and Washington, and a national-level evaluation conducted in 12 states that was authorized by the Safe, Accountable, Flexible, Efficient Transportation Equity Act: A Legacy for Users (SAFETEA-LU).[9] We also reviewed preliminary results from two ongoing pilot projects in Minnesota and Nevada. We interviewed transportation officials involved with all of these pilots. To gather information on the benefits and challenges of international mileage fee initiatives, we selected programs in Germany, New Zealand, and the Netherlands based on criteria that included program objectives, administration, and technologies used. We reviewed program documents and interviewed transportation officials in these countries to gather information on Germany's Heavy Goods Vehicle user fee system, New Zealand's Road User Charge system, and a cancelled user fee program in the Netherlands. To determine the mileage fee rates that would be necessary to replace and supplement current Highway Trust Fund revenues, we selected three revenue targets and simulated the user fee rates for passenger vehicles and commercial trucks that would be necessary to achieve those targets using fiscal year 2010 data. We also simulated the mileage fees that would be required if all vehicles were responsible for the road damage they caused according to Federal Highway Administration (FHWA) estimates. To obtain states' views on addressing future revenue demands using mileage fees, we surveyed the DOTs in all 50 states and the District of Columbia and received a 100 percent response rate.

We conducted this performance audit from December 2011 through December 2012 in accordance with generally accepted government auditing standards. Those standards require that we plan and perform the audit to obtain sufficient, appropriate evidence to provide a reasonable basis for our findings and conclusions based on our audit objectives. We believe that the evidence obtained provides a reasonable basis for our findings and conclusions based on our audit objectives. Our objectives, scope, and methodology are discussed in more detail in appendix I.

BACKGROUND

Financing of U.S. Surface Transportation Programs

Since Congress established the Highway Trust Fund in 1956 to fund the construction of the Interstate Highway System, the federal government has financed transportation projects primarily with revenues collected through federal fuel taxes. These taxes were established to make the federal-aid highway program self-financing—that is, paid for by the highway users who directly benefit from the program.[10] From 1956 to 1993, Congress increased the federal gasoline tax from 3 cents per gallon to its current rate of 18.4 cents per gallon. The federal diesel tax rate was also last increased in 1993 and is currently set at 24.4 cents per gallon. In fiscal year 2010, 92.5 percent of almost $37 billion in user fee revenues deposited into the Highway Trust Fund were generated through federal fuel taxes.[11] The remaining revenues were collected through several taxes on heavy commercial trucks.[12] (See fig. 1.)

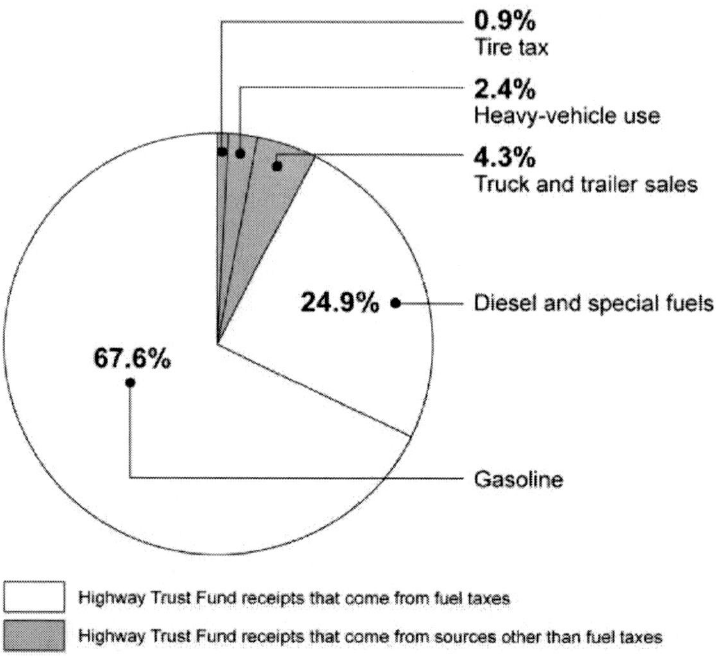

Source: GAO analysis of FHWA data.

Figure 1. Sources of Revenue for the Highway Trust Fund, Fiscal Year 2010.

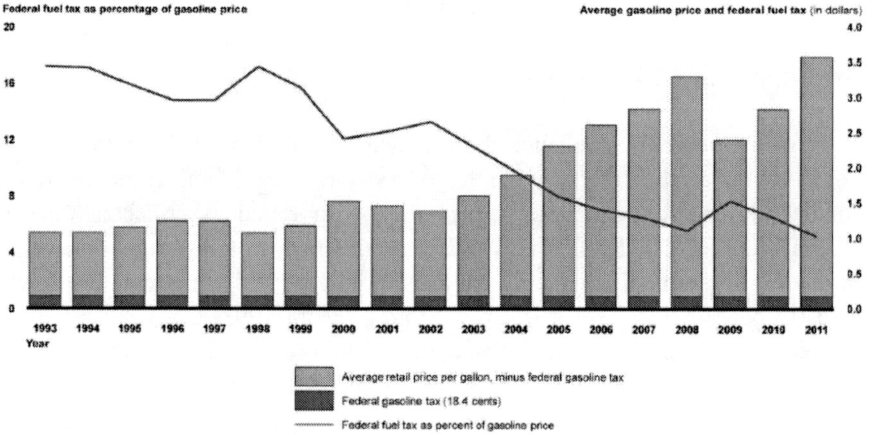

Source: GAO analysis of Department of Energy data.

Figure 2. Federal Gasoline Tax as a Percentage of the Average Retail Price per Gallon of Gasoline, 1993 to 2011.

Over the past two decades, revenues from federal gasoline and diesel fuel taxes have steadily declined in purchasing power. These rates are not set to be adjusted for inflation on an annual basis, meaning that the 18.4 cents per-gallon tax on gasoline enacted in 1993 is effectively worth about 11.5 cents today.[13] If federal gasoline taxes had been indexed to inflation since they were last increased, they would have risen from 18.4 cents per gallon in 1993 to approximately 29 cents per gallon in 2011; diesel taxes would have risen from 24.4 cents per gallon to 38 cents per gallon. In addition, as shown in figure 2, the amount of federal gasoline tax relative to the average retail price of gasoline decreased from 17 percent in 1993 to 5 percent in 2011.[14]

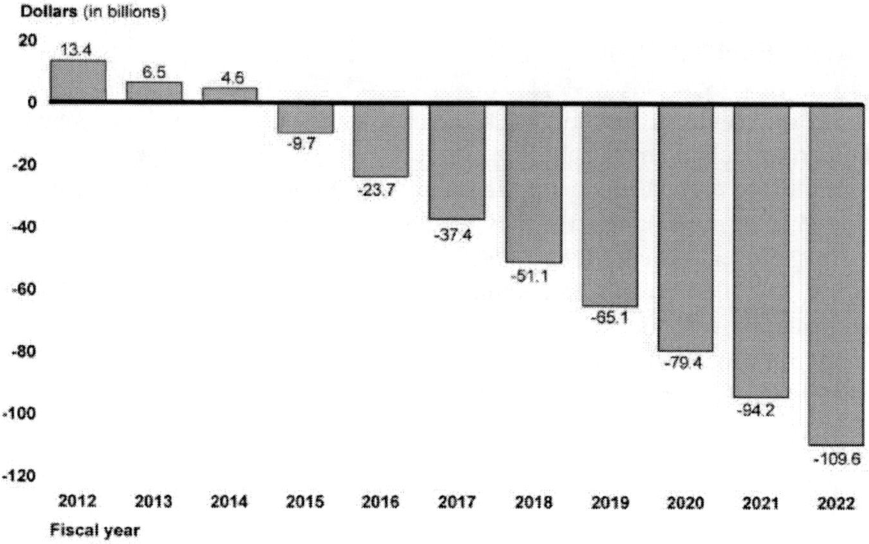

Source: GAO analysis of CBO data.

Note: This projection assumes no further appropriations after 2014 from general revenues to the Highway Trust Fund.

Figure 3. Projected Highway Trust Fund Balance, 2012 to 2022.

Surface transportation programs face increasing shortfalls in year-to-year revenues over the next decade. CBO estimated in August 2012 that, to maintain current spending levels from 2012 to 2022, the Highway Trust Fund would require an additional $110 billion over what it is expected to take in during that period (see fig. 3). These shortfalls are expected to increase as passenger vehicle fuel economy improves and the amount of revenue generated per mile traveled decreases. When federal fuel taxes were last increased in 1993, the National Highway Traffic Safety Administration's Corporate Average Fuel Economy (CAFE) standards required newly manufactured passenger cars and light trucks to meet estimated average fuel economy levels of 27.5 miles per gallon (mpg) and 20.4 mpg, respectively. Current CAFE standards will require manufacturers' new passenger cars and light trucks to have an estimated combined average fuel economy of 34.1 mpg by model year 2016,[15] and as high as 54.5 mpg by 2025.[16] We have previously reported that, over the long term, vehicles will become more fuel efficient and increasingly run on alternative fuels. Consequently, fuel taxes may not be a long-term source of transportation funding.[17]

Users' Fuel Tax Costs

The amount of federal fuel taxes paid by drivers of passenger vehicles comprises only a small portion of these users' overall fuel expenditures and varies based on fuel economy. More fuel-efficient passenger vehicles pay less in fuel taxes per mile because they require less fuel to travel the same distance as less efficient vehicles. The *2009 National Household Transportation Survey* found that the average one-vehicle household spends more than $1,400 annually on gasoline.[18] Drivers of sedans with average fuel efficiency (22 mpg) pay about $100 per year in federal fuel taxes. The driver of a sport utility vehicle (SUV) with a fuel efficiency of 16 mpg pays about $132 a year in federal fuel taxes (see fig. 4).

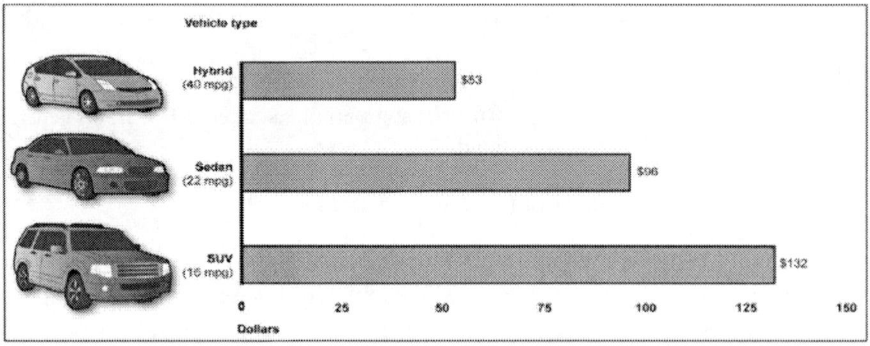

Source: GAO analysis of FHWA and Department of Energy data.
Note: Assuming 11,489 miles driven annually (per FHWA 2010 Highway Statistics).

Figure 4. Average Annual Federal Fuel Taxes Paid by Passenger Vehicles.

A single commercial truck contributes more to the Highway Trust Fund, on average, than a single passenger vehicle for several reasons. The federal diesel fuel tax is higher than the gasoline tax, and commercial trucks tend to travel more miles annually and are considerably less fuel efficient than passenger vehicles. According to FHWA highway statistics, the average commercial truck and trailer combination (combination truck) traveled nearly 70,000 miles in 2010 while the average passenger vehicle traveled about 11,000 miles. Also, while the average combination truck traveled about 6 miles per gallon of diesel fuel in 2010, the average passenger vehicle traveled about 22 miles per gallon of gasoline. There is also a great deal of variation in the amount of fuel taxes paid among different configurations of commercial trucks. As shown in figure 5, a combination truck with a fuel efficiency of 5 mpg would have paid more than 8 times as much in federal diesel fuel tax as a single-unit truck with a fuel efficiency of 8 mpg, a difference due primarily to combination trucks driving substantially more miles than smaller, single-unit trucks.[19]

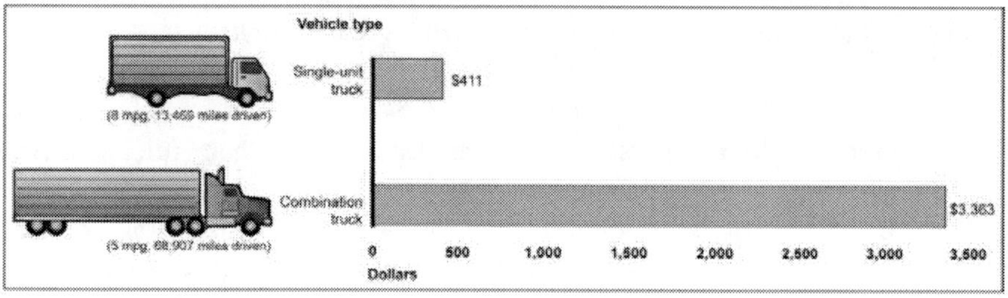

Source: GAO analysis of FHWA data.

Figure 5. Annual Federal Fuel Taxes Paid by Commercial Trucks in 2010.

Although a single commercial truck generally contributes larger amounts to the Highway Trust Fund through federal diesel fuel and other taxes, FHWA has estimated that commercial trucks pay less than their share for their use of our nation's roadways in relation to the road damage they impose. According to research conducted by the American Association of State Highway Transportation Officials (AASHTO), highway wear increases exponentially with the weight of a vehicle's axle load.[20] To evaluate the equity and efficiency of highway user fees, FHWA has compared the Highway Trust Fund contributions of different vehicle classes with the costs attributable to each class as part of its *Highway Cost Allocation Study*. FHWA completed its most recent cost allocation study in 1997 in response to a GAO recommendation and provided an updated addendum in 2000.[21] This study found that in general, lighter vehicles pay more than their share of highway costs while heavier vehicles pay less than their share. For example, according to FHWA's 2000 study, the heaviest combination trucks—those that weigh more than the federal interstate gross vehicle weight limit of 80,000 pounds—paid 50 cents for every dollar's worth of damage they caused.[22] According to the AASHTO study that is used as the basis for the organization's pavement design guides, a commercial truck with five axles weighing 80,000 pounds imposes roadway damage equivalent to the damage imposed by 24,000 passenger cars.

Mileage-based User Fee Initiatives in the United States and Abroad

In recent years, the federal government and several states have taken steps to evaluate mileage fee systems, although none of these U.S.- based pilot projects has collected fees from drivers based on their road use:

- In 2005, Congress authorized $16.5 million for a field test for assessing highway use fees to vehicles based on their mileage driven and using satellite-enabled, on-board units.[23] Led by researchers at the University of Iowa, the National Evaluation of Mileage-based Road User Charges (referred to as the "Iowa study" in this report) tested mileage fee systems in the vehicles of 2,600 volunteer study participants in 12 states.[24]
- In 2005 and 2006, respectively, the Puget Sound Regional Council—the metropolitan planning organization for the Seattle, Washington region—and the Oregon DOT each conducted pilot programs that were funded in part by FHWA's Value Pricing Pilot Program.[25] Similar to the Iowa study, both pilots tested GPS-based systems installed into the vehicles of paid volunteers to gather mileage data and calculate hypothetical user fees.

At the time of our review, the Minnesota and Nevada DOTs were testing different approaches to calculating mileage fees for passenger vehicles, discussed later in this report. The Oregon DOT also conducted field tests of a system to collect mileage fees from commercial trucks and has planned a new pilot project to examine mileage fees for electric passenger vehicles. For a summary of U.S.-based pilot projects, see appendix II.

Internationally, several countries have implemented distance-based user fee programs for commercial trucks, including Germany, New Zealand, Switzerland, Austria, the Czech Republic, and Slovakia.[26] These programs were designed to address a variety of policy goals, including raising revenues and reducing harmful emissions. The New Zealand system also charges diesel-fueled passenger vehicles for distances traveled. The Netherlands attempted to implement a nationwide commercial truck and passenger vehicle mileage fee system scheduled to begin in 2012, but the program was suspended before it was implemented because of privacy concerns discussed later in this report.

Evaluating User Fees

Prior GAO work has found that the design of user fee programs can be evaluated based on several related criteria: efficiency, equity, revenue adequacy, and administrative burden, as discussed in appendix III.[27] These criteria interact and are often in conflict with each other; as such, there are trade-offs to consider among the criteria when designing a fee. For example, the current method of collecting federal fuel taxes presents little administrative burden because these taxes are collected from a small number of companies that store or distribute fuel at the wholesale level.[28] However, according to CBO, fuel taxes also raise efficiency and equity concerns in that they: (1) do little to promote the efficient use of the nation's roadways because they provide minimal incentive for users to drive less and (2) tend to be regressive, in that they impose a larger relative burden on low-income than on high-income households.[29] Consequently, every user fee design will have pluses and minuses, and no design will satisfy everyone on all dimensions.

The implementation of mileage fee systems may also be evaluated against several widely accepted principles for protecting the privacy and security of personal information. These principles, known as the Fair Information Practices, include collection limitation, data quality, purpose specification, use limitation, security safeguards, openness, individual participation, and accountability.[30] In prior work, we noted that these principles, with some variation, are used by organizations to address privacy considerations in their business practices and are also the basis of privacy laws and related policies in the United States, the European Union, Australia, and New Zealand.[31]

Various Initiatives Demonstrate Mileage Fee Benefits, but Privacy Concerns and Cost Challenges Hinder Passenger Vehicle Systems

Mileage fee initiatives in the United States and abroad show that several approaches are available to gather mileage data and charge fees. Some approaches could lead to more equitable and efficient use of roadways; however, significant privacy-related concerns from the public and cost challenges have been raised in applying mileage fees to passenger vehicles. Reliable cost estimates for mileage fee systems are not available; but launching and operating a system to collect fees from 230 million U.S. passenger vehicles is expected to greatly exceed the current costs of collecting federal fuel taxes. Commercial truck user fee systems in Germany and New Zealand show that considerable revenues and other benefits can be achieved by charging these vehicles, but enforcing compliance in a cost-effective manner presents trade-offs. Only limited research has been done to evaluate commercial truck mileage fees in the United States. Recent efforts in two states suggest that charging mileage fees to commercial trucks presents several benefits over passenger vehicle fees, including fewer privacy-related concerns and cost challenges.

Several Different Passenger Vehicle Mileage Fee Approaches Exist

The five U.S.-based pilot projects that we reviewed and New Zealand's passenger vehicle mileage fee program illustrate three general approaches that are currently available to gather mileage data and charge drivers user fees.[32] The approaches vary in terms of the specificity of the mileage data collected as well as the procedures used to charge drivers fees. The three approaches are a GPS-based system; a pay-atthe-pump system; and a prepaid, manual system, as depicted in figure 6 and described in more detail below.

GPS-based Mileage Fee Systems

To implement mileage fees using GPS, the vehicle must be installed with an on-board unit with a GPS chip to receive signals from a satellite to determine the vehicle's location. Two types of GPS-based systems can be used to calculate mileage fees. One system has sufficient processing capability to calculate mileage fees in the vehicle (known as a "thick client"); the other system (known as a "thin client") has less processing capability and sends the location data to a central office where mileage fees are calculated. Three of the five U.S.-based pilot projects we reviewed tested on-board thick-client units, one used a thin-client unit, and one did not use GPS.[33] Thick-client systems transmit summary information on the total miles driven by jurisdiction and the amount due to a central office, which then prepares the participant's mileage fee invoice.[34] Thin-client systems work similarly, but rather than sending a summary of the mileage traveled to a central office, they send detailed vehicle location and time-of-day data. The central office uses the data to calculate and prepare the participant's mileage fee invoices.

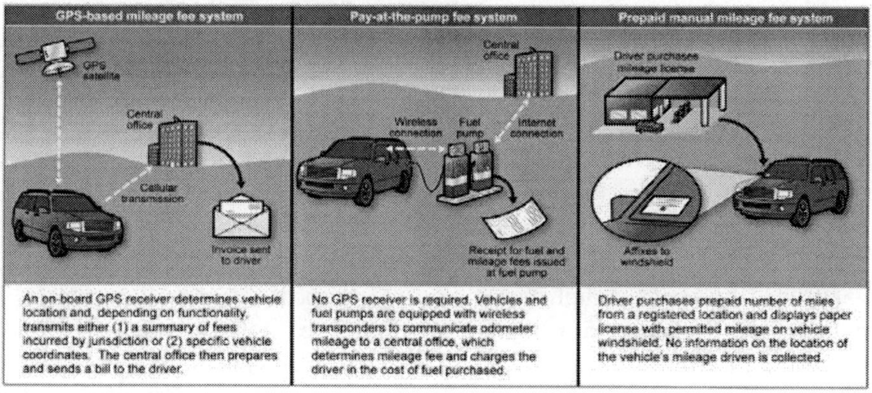

Source: GAO analysis of mileage fee initiatives.

Figure 6. Depiction of Three Passenger Vehicle Mileage Fee Approaches.

Pay-at-the-Pump Mileage Fee Systems

Two states, Nevada and Oregon, have designed mileage fee collection systems that resemble the way in which fuel taxes are currently paid—at the filling station. The Nevada DOT is currently conducting a pilot project that does not require an on-board GPS system and that estimates a driver's mileage fee based on the fuel efficiency of the vehicle and the amount of fuel purchased at the pump. To do so, a wireless transponder is installed in the participant's vehicle and connected to the vehicle's onboard diagnostics unit to gather the total mileage of the vehicle. When a vehicle pulls up to a fuel pump at a gas station participating in the pilot, the transponder in the vehicle sends the vehicle's mileage information to a transponder installed at the fuel pump. The vehicle's mileage information is then transmitted to a central office where a mileage fee is calculated and transmitted back for inclusion in the price of fuel and shown on the participant's fuel receipt. In 2006, the Oregon DOT tested a similar pay-at-the pump system, but its system also required a GPS receiver [35] in participant vehicles to charge drivers different fee rates based on the jurisdiction of travel (in state, out of state, or within the Portland metropolitan area).[36]

Prepaid Manual Mileage Fee Systems

A prepaid manual system is a nonautomated mileage fee system where drivers purchase a license that permits them to drive for the purchased number of miles, as used in New Zealand's Road User Charge (RUC) system. The RUC system was originally designed to collect user fees from commercial trucks, but it also applies to diesel-fueled passenger vehicles, which comprise 15 percent of all light vehicles in the country. New Zealand does not tax diesel fuel at the pump because a significant amount is used for off-road purposes that do not impose costs on public highways—such as agriculture, construction, fishing, and logging. Vehicle owners purchase a RUC license for a range of kilometers (such as 5,000 to 10,000 kilometers) based on their odometer reading, and the RUC license shows the distance the vehicle is permitted to drive. Vehicle owners must display the RUC license on the windshield to verify compliance during safety inspections or if a vehicle is pulled over for traffic a violation. RUC licenses are purchased in 1,000-kilometer increments at a variety of locations including post offices, some filling stations, and through the New Zealand Transport Agency, which manages the program. Unlike the pay-at-the-pump system, New Zealand's RUC system can be used to collect mileage fees from electric and alternative fueled vehicles that do not visit fuel pumps but do use roads.[37] The RUC system was the only passenger vehicle system we reviewed that collects actual revenues and the only national mileage fee system that applies to passenger vehicles.

Benefits and Challenges of Passenger Vehicle Mileage Fee Systems

The three approaches offer different benefits and challenges and present trade-offs to policymakers in evaluating mileage-based user fees. GPSbased systems can lead to more equitable and efficient use of roadways by charging drivers based on their actual road use and by providing pricing incentives to reduce road use. However, public perception of privacy risks raised by these systems significantly limits acceptance of the program. Conversely, because pay-at-the-pump and prepaid manual systems do not collect location data on drivers they present fewer privacy-related challenges, but the trade-offs are reductions in the efficiency and equity of the proposed systems. For example, they are unable to improve the efficiency of road use by charging drivers different rates for travel on specific roadways or during congested periods. Manual systems could also be subject to odometer fraud and evasion, with compliant drivers paying more than noncompliant drivers. Because reliable cost estimates for implementing any of these systems are not available, the ability to weigh the costs of implementing and operating a system is a challenge that applies across each of the options. See table 1 for a summary of the benefits and challenges.

Table 1. Benefits and Challenges of Mileage Fee Systems

Mileage fee systems	Benefits	Challenges
GPS systems	Opportunity to improve the efficiency of road use through variable pricing, or charging road users a higher rate during peak traffic times and a lower rate during times with light traffic. Ability to improve equity by accurately pricing road use for all users and vehicle types consistent with the costs imposed.	Perception of government intrusion on privacy by tracking privately owned vehicles and risk of data beingcompromised or disclosed for unauthorized uses. High start-up and ongoing costs associated with unit installation, implementation, billing, revenue collection, and enforcement.
Pay-at-the-pump systems	Alleviates some privacy-related concerns of tracking privately owned vehicles by not requiring the use of a GPS system.	Cost and logistical challenges associated with the installation and management of equipment at gas stations nationwide and installation of transponders in vehicles. Unable to gather driving data needed to implement variable pricing based on congestion to encourage efficient road use. Not compatible with alternative fuel vehicles that do notuse gas stations.
Prepaid manual systems	Alleviates privacy-related concerns of tracking privately owned vehicles by not requiring the use of GPS system. Limits administrative costs by eliminating the need to outfit vehicles or fuel pumps with the necessary technology to collect total miles traveled.	Unable to gather driving data needed to implement variable pricing to encourage efficient road use. Increased risk of evasion through odometer tampering creates equity issues.

Source: GAO analysis.

Privacy Challenges

Using a GPS device to charge passenger-vehicles mileage fees raises significant privacy-related concerns among the public. In the surveys conducted through the Iowa study, researchers found that after completing the pilot, 60 percent of the volunteer participants believed that the government would use the information collected to track their movements. In addition, state DOT officials conducting public outreach and opinion research in Minnesota, Texas, and Nevada found that the public expressed a variety of privacy-related concerns, including that the government would use a GPS system to track a driver's movements and that the personal information collected could be vulnerable to security breaches or shared with law enforcement agencies and private companies. According to our survey of state DOTs, 45 of 51 officials reported that addressing privacy-related concerns would present a great challenge to developing a mileage fee program in their state. Several state DOT officials have proposed that using the private sector to manage mileage fee programs would alleviate public perception of privacy risks by limiting the government's role in collecting and managing personal information. However, we have previously reported that by allowing private companies access to location data, users can be exposed to privacy risks, including disclosure of the location data to unknown third parties for unspecified uses, consumer tracking, identity theft, threats to physical safety, and surveillance.[38]

Because of public perception of privacy risks, Oregon and Nevada adopted an approach that does not require the use of GPS-based systems in future pilots and the Netherlands suspended its mileage fee initiative. Oregon DOT officials stated that a government-mandated GPS device installed in a vehicle to track and charge for the vehicle miles traveled is no longer feasible because of public privacy-related concerns. As a result, the Oregon DOT is conducting an alternative pilot program that does not mandate the use of a GPS device and provides users several options of reporting and paying for miles traveled.[39] In designing its pilot project, the Nevada DOT received negative public feedback over the use of GPS devices to determine miles driven. As a result, Nevada DOT is testing a pay-at-the-pump system which, according to officials, alleviates the public's privacy-related concerns. Similarly, a commercial truck and passenger vehicle user fee system that was scheduled to begin in 2012 was suspended by the Dutch government before it could be implemented. Dutch officials stated that negative media coverage of the program's potential to invade personal privacy led to public uncertainty about the program and subsequent opposition from a number of political parties. Consequently, the government deemed the user fee program too controversial and suspended it. Furthermore, concerns with mileage fee systems have informed congressional debate. In June 2012, the House of Representatives included a provision in its fiscal year 2013 appropriations bill for the U.S. Department of Transportation that would have prevented the Secretary of Transportation from using any of the funds made available under that legislation to research or implement a system that would levy a fee on a vehicle user based on the distance traveled.[40]

Pay-at-the-pump or prepaid manual systems address privacy-related concerns to some extent by not tracking a driver's location; however, these systems can reduce some of the efficiency and equity benefits of mileage fee systems. For example, variable pricing programs that charge road users a higher rate during peak traffic times and a lower rate during times with light traffic can be facilitated through the use of GPS systems. In theory, GPS could also be used to charge drivers different rates based on the type of road or the location of travel, such as higher rates on urban Interstates and lower rates on rural roads. The system in place in New Zealand and the one being tested in Nevada only collect total mileage driven by the vehicle and are unable to apply different fees based on time and location. In addition, pay-at-the pump systems cannot charge electric fuel vehicles since they do not need to use gas stations to power their vehicles. Also, equity issues caused by odometer fraud exist in prepaid manual systems. For example, New Zealand officials reported that when a passenger vehicle is pulled over for a violation, the police have no way to determine whether an odometer has been disabled and that mileage is not being recorded. An independent review of the RUC program conducted in 2008 estimated that about $10 million (U.S.), or 6 percent of the light vehicle RUC revenue, was lost because of evasion.[41] New Zealand government officials noted that as a result, compliant drivers pay higher RUC rates than they otherwise would to replace the revenues lost from noncompliant drivers.

Cost Challenges

Reliable estimates for start-up and ongoing administrative costs of a passenger vehicle mileage fee system in the United States are currently unknown but are likely to be substantial.

- *Start-up costs:* Installing on-board units in 230 million U.S. passenger vehicles would almost certainly be a significant cost challenge. Reliable estimates for the current cost of purchasing and installing on-board units for all U.S. passenger vehicles are not available, but such costs are likely to greatly exceed the current costs of collecting fuel taxes, estimated at about 1 percent of the revenues collected. A May 2009 study, prepared for U.S. Department of Transportation on the administrative costs of collecting highway revenues, found that the costs of purchasing and implementing a GPS-based mileage fee system could range from roughly 8 percent to 33 percent of the revenues generated over a 20-year period, depending on the type of system used. The German government estimated the current cost of the GPS-enabled on-board units used in its commercial truck user fee system to be about $240 per unit, with additional costs for installation.[42] These start-up costs would likely be prohibitive if the goal of the fees is to replace current federal fuel tax revenues, which are about $100 per year for the driver of a sedan with average fuel efficiency. Similarly, retrofitting thousands of gas stations to support a pay-at-the pump system would be costly and challenging. For example, the Oregon and Nevada pay-at-the-pump pilot programs cited difficulties finding and recruiting gas stations to participate in their pilot programs.

- *Ongoing administrative costs:* The costs of managing, maintaining, and enforcing any mileage fee system are also unknown but likely to be substantial. For example, a 2011 report analyzing cost estimates from the proposed mileage fee system in the Netherlands estimated the total operating costs to be about 7 percent of the revenues projected to be generated by that system.[43] However, the report found that operating costs as a percentage of revenues could be substantially higher in the United States because U.S. drivers pay substantially less in transportation costs than drivers in the Netherlands. Furthermore, the ongoing maintenance of on-board units could be costly and present technical challenges. In the Iowa study, approximately 24 percent of the 2,600 participants experienced at least one study-related problem with their vehicle or installed equipment over the 2-year study period, and a total of 618 incidents required at least one service visit to correct. Of these incidents, 79 percent were due to a problem with the performance of installed onboard units. The principal researcher on the project stated that this level of error would have been "disastrous" and would have jeopardized the success of a national mileage fee rollout with millions of vehicles involved. The Iowa study recommended that the federal government should fund a larger, national mileage fee study that would generate the efficiencies needed to provide an accurate estimate for back-office operational costs.

Commercial Truck User Fee Programs in Germany and New Zealand Have Raised Substantial Transportation Revenues and Produced Other Benefits

Revenues and Benefits from Germany's Heavy Goods Vehicle User Fee System

Germany's Heavy Goods Vehicle (HGV) system was implemented in January 2005 and is the first distance-based user fee system in Europe to use GPS technology.[44] The program was developed through a public-private partnership between the German government and a private sector company, which manages and collects tolls for the German government. The HGV system charges all trucks weighing more than 12,000 kilograms (over 26,000 pounds), regardless of national origin, a per-kilometer fee to travel on the 12,700-kilometer (almost 8,000-mile) national motorway, or autobahn. Currently, about 700,000 trucks across Europe are equipped with GPS-based on-board units that meter all travel on the autobahn. The on-board units use cellular communications to transmit mileage data to the private toll operator, which is responsible for billing trucking firms and collecting the fees. Approximately 700,000 other commercial trucks without on-board units are charged user fees through a manual booking system maintained by the private toll operator. Users of this system must prepay for their travel through the Internet or at 3,500 electronic toll terminals located at various entry points and service areas across the autobahn.

The goals of the HGV system were to (1) raise transportation revenues through the user pays principle and to (2) reduce harmful emissions from commercial trucks. The program has generally achieved its goal of raising new revenues by ensuring that both foreign and domestic trucks on the autobahn are charged equitably for their travel. According to German transportation officials, foreign trucks account for more than 35 percent of the country's truck travel. Prior to the launch of the HGV system, foreign trucks could purchase diesel fuel in neighboring countries to avoid Germany's significant fuel taxes, and thus use German roadways without contributing revenues for their maintenance. The HGV tolling system addresses this by charging all trucks subject to the tolls, regardless of national origin, the same per-kilometer fees to travel on the autobahn. From fiscal year 2007 through 2011, the HGV system raised almost $25 billion (more than 20 billion euros) in dedicated surface transportation revenues from an estimated 1.4 million commercial truck users.[45]

The program also achieved its second goal of creating incentives for operators to invest in lower emission vehicles, which has resulted in reduced emissions across the German and European trucking fleet. User fee rates are variable and based on the truck's emission class, number of axles, and distance traveled on the roads subject to the fee. Trucks with lower emissions pay significantly less than trucks with higher emissions, with rates ranging from 28 to 57 U.S. cents per mile. The composition of the HGV trucking fleet has changed dramatically over the course of the program as a result of the variable pricing incentives. In 2005, the lowest emission commercial truck classes, known as the Euro 5 and enhanced environmentally-friendly vehicle emissions categories, comprised less than 1 percent of the commercial trucking fleet. By the end of 2011, those classes comprised about 70 percent of the commercial truck fleet. German transportation officials reported that while this outcome was a positive environmental benefit, the fast turnover of the vehicle fleet resulted in slightly lower-than-expected revenues because lower emission vehicles pay lower rates. Consequently, the German government reviewed and raised the rates across all categories while maintaining the incentives for lower emission vehicles.

Revenues and Benefits from New Zealand's RUC Program

The New Zealand RUC program was established in 1977 and requires that owners of vehicles over 3,500 kilograms (7,700 pounds), including trucks and their trailers, prepurchase a RUC license for all miles traveled on the nation's 94,000-kilometer (58,000-mile) public roadway system. As with diesel passenger vehicles, commercial trucks purchase RUC licenses in 1,000 kilometer increments and must display the license and the kilometers permitted on their windshield. The government has provided an option for vehicle owners to install electronic RUC (E-RUC) systems that are provided by private sector firms certified by the New Zealand government to collect fees and remit revenues to the government. The E-RUC system uses on-board GPS units to keep track of the status of each vehicle's RUC license and whether additional kilometers need to be purchased. Trucking firms that subscribe to the system can track their fleet's RUC licenses online and may purchase additional applications enabled by GPS to track the location, speed, and fuel efficiency of their trucks.

Commercial trucks, rather than passenger vehicles, are the vast majority of E-RUC users because of the higher cost of subscribing to E-RUC systems compared with lower costs of purchasing prepaid paper RUC licenses directly from retailers.

The primary goal of the RUC program is to generate surface transportation revenues from heavy vehicles by charging users fees that reflect the maintenance costs that these vehicles impose on the roadways. Because highway wear increases exponentially with the weight of a vehicle's axle load, the RUC system uses variable pricing to charge higher rates to heavy trucks with fewer axles, and lower rates to heavy trucks with more axles. For example, a three-axle truck weighing more than 18 tons (almost 40,000 pounds) is charged almost 43 U.S. cents per mile, while a truck of the same weight with five or more axles is charged 35 U.S. cents per mile.[46] All miles for light passenger diesel vehicles are charged at a flat rate of less than 6 U.S. cents per mile regardless of the actual vehicle weight.

This is because an individual light vehicle causes only very limited amounts of wear on the road compared with heavy trucks, and the variation in wear among light vehicles of different weights is also minimal. From 2007 through 2011, the RUC system generated about $3.4 billion (U.S.) in dedicated surface transportation revenues, with 74 percent coming from heavy trucks and 26 percent from light passenger vehicles.[47]

The New Zealand RUC program has achieved a related program benefit of promoting the use of trucks that cause less damage to roadways, an outcome that may reduce road maintenance costs. According to New Zealand government officials, the variable pricing incentives have influenced the composition of the commercial truck fleet so that trucks with more axles per vehicle weight represent a greater portion of the vehicle fleet than in most other countries.

For example, according to 2010 New Zealand Transport Agency data, combination trucks with eight axles represented about 35 percent of all trucks on New Zealand roadways. In contrast, according to FHWA data, eight-axle combination trucks were estimated to account for less than 1 percent of the U.S. truck fleet in 2000.[48] New Zealand trucking association representatives reported that operators in New Zealand use trucks and trailers with as many as nine axles in combination in order to pay lower RUC fees. New Zealand government officials reported that this altered vehicle fleet is considered to reduce road maintenance costs due to reduced pavement damage, although no estimate is available for the overall value of the savings.

Commercial Truck Mileage Fee Systems Face Trade-Offs between Cost and Enforcement While Acting to Mitigate Privacy Concerns

The different approaches taken by Germany and New Zealand illustrate the trade-offs involved in designing a mileage fee system for trucks that provides adequate enforcement in a cost-effective manner. The German system employs an extensive but costly roadside enforcement infrastructure; New Zealand's RUC program does not, but it has higher estimated evasion rates and revenue leakages than the German system.

The German government and its private sector toll operator ensure compliance with the HGV system through significant roadside infrastructure and enforcement. The HGV system includes 300 gantries, or overhead structures, deployed across the autobahn to ensure that onboard units are operating correctly and that manual users of the system have registered for their travel. Each gantry is equipped with enforcement cameras and short-range, wireless devices that signal to each passing truck's on-board unit to ensure it is activated and functioning properly.[49] The German government staffs a fleet of 250 mobile enforcement vehicles equipped by the toll operator, which is used to patrol the autobahn and respond to suspected violations.[50] (See fig. 7.) According to officials, this approach has resulted in an estimated evasion rate that is less than 1 percent, meaning that less than 1 percent of all HGVs on the system were out of compliance in 2011.

Source: German Ministry of Transport and Urban Development and the Federal Office for Goods Transport.

Figure 7. German HGV Enforcement Gantry and Mobile Enforcement Vehicle.

Although the German government reported that the HGV system experiences very little revenue leakage from evasion, the system is costly to implement. All of the start-up costs for implementing the HGV system were incurred by the toll operator, which owns the 700,000 on-board units installed across Europe, as well the roadside infrastructure and back-office equipment used to monitor and enforce the system.[51] From fiscal years 2007 through 2011, the German government paid the system operator approximately $664 million per year to manage the system. These administrative costs averaged about 13 percent of the $5 billion in average revenues generated annually over that time period. The German government also spends almost $62 million per year to staff the mobile enforcement vehicles with German police. In addition, Germany allocates $740 million annually in HGV revenues for several programs to assist German trucking firms in complying with the system. Although these compliance programs are not considered to be part of the administrative costs of operating the HGV system, they further reduce the net revenues generated.[52]

New Zealand's RUC fees are collected and enforced manually with little roadside enforcement technology and lower costs than the HGV system, but with higher estimated evasion rates. The New Zealand national police enforce RUC compliance with eight inspection stations across the country, staffed with a total of 90 officers nationwide. Commercial trucks are required to stop at these stations, and police manually inspect and compare the mileage limit on the displayed RUC license with the vehicles' current mileage on special odometers that are mounted on the hubs of trucks and trailers.[53] According to the New Zealand trucking association, compliance with these manual inspections is inefficient and costly to operators because of time lost. The government reported that the costs of managing the RUC program are nearly $18 million (U.S.) per year, or 2.5 percent of the estimated $700 million (U.S.) generated annually by the --system.[54] Although these administrative costs are substantially lower than the 13 percent costs of the German system, implementing the RUC system comes with higher estimated evasion rates than the 1 percent of revenues estimated lost in Germany. According to the New Zealand Ministry of Transport, roughly 4 percent of the commercial truck revenues on average are lost annually because of evasion by trucks. This equates to an estimated loss of about $21 million (U.S.) per year from fiscal years 2007 to 2011, or slightly more than the amount that the New Zealand government spends annually to manage the program.[55]

Germany and New Zealand officials reported they have safeguards in place to limit the collection and use of mileage data, and as a result, privacy concerns for commercial trucks are not a significant challenge. The officials reported that the respective laws that authorize these programs stipulate several data protection provisions consistent with internationally recognized Fair Information Practices.[56] Both programs are required to clearly define and limit the data to be collected, the purpose for its collection, the limits for its use, as well as the security safeguards that are in place. For example, in Germany, the government and its contracted toll operators are permitted to collect, use, and process only limited information to enforce the tolls, including a picture of the vehicle, the place and time the tolled road was used, and features of the vehicle (number of axles) necessary to process the toll. This data must be deleted immediately after the toll is paid. Because of the legal safeguards in place, commercial trucking associations in Germany and New Zealand both reported that the invasion of personal privacy or the loss of proprietary business information are not significant concerns among their firms. Because the New Zealand RUC system is largely a prepaid manual system in which most users do not have an on-board unit, privacy-related concerns do not arise. Moreover, safeguards are in place to protect the data of the commercial truck fleet using the E-RUC system. According to New Zealand government officials, the private companies applying to become E-RUC providers are subjected to a stringent series of tests to ensure the security of data before the government approves them to act as agents for the RUC system. Tests include verification of the providers' data security systems, the accuracy of the data collected, and their systems' ability to identify efforts to tamper with E-RUC units.

Commercial Truck Mileage Fee Pilot Projects in the United States Also Suggest Benefits

Although U.S. initiatives to evaluate mileage fees have focused primarily on passenger vehicles, two recent efforts focused on fees for commercial trucks. Specifically, the Oregon DOT and several metropolitan planning organizations and a consulting firm in New York recently evaluated ways to electronically collect existing weight-distance taxes that apply to commercial trucks in those states.[57]

- In 2010, the Oregon Truck Road Use Electronics (TRUE) pilot project tested GPS-based units in 25 trucks operated by three trucking firms to automate the collection of Oregon's truck weight-mile tax. Oregon officials reported that the devices successfully tracked the miles traveled in their state and sent the data to Oregon DOT to produce a monthly weight-mile tax statement for the trucking firms participating in the pilot, to facilitate their payment of the state's weight-mile tax.

- In a separate 2010 study, researchers collected GPS routing data provided by several trucking firms in New York to test the feasibility of using existing GPS-based technology to implement a truck mileage fee system that would replace existing state truck fees and taxes. [58] The study found that a mileage fee system could more accurately track vehicle mileage than the current system, which relies on self-reported mileage from trucking firms, and could generate an additional $150 million in revenue annually that is estimated to be lost because of underreporting of mileage and weight.

The results of these U.S. pilots suggest that charging mileage-based user fees to commercial trucks presents several benefits over passenger vehicle fees:

- *Reduced privacy concerns.* Some of the privacy-related concerns discussed previously with regard to using GPS-based systems for passenger vehicles may not be as troublesome if applied to commercial trucks because the right of privacy would belong to the company that owns the truck. Trucking companies can be required to meet reasonable conditions and pay appropriate taxes in exchange for using the public highways. They can also set the conditions of employment for their employees, who have limited privacy rights because employers may establish the policies that govern their workplaces and allow the monitoring of the use of any company assets, including trucks. Employers may also establish policies that inform employees about information collected as a condition of employment. The New York commercial truck mileage fee study reported that a significant portion of large trucks already have fleet management systems that include the GPS technology needed to support mileage fees. These fleet management systems are used by trucking firms to monitor mileage traveled, driver speed, estimated delivery times, and other business performance information.
- *Reduced implementation costs.* First, because there are significantly fewer commercial trucks than passenger vehicles in the United States, the overall costs of implementing a truck mileage fee system would almost certainly be significantly lower for these vehicles. According to FHWA data, commercial trucks represent 4 percent of the U.S. vehicle fleet, with a total of about 10 million single-unit and combination vehicles on the roadways in fiscal year 2010. Equipping

trucks with on-board units would be significantly less costly than equipping the 230 million passenger vehicles with the same equipment. In addition, because combination commercial trucks average significantly more miles per year (almost 70,000) than the average passenger vehicle (more than 11,000 miles), commercial truck mileage fee systems could generate significantly more revenue per vehicle than passenger vehicles systems. As such, the costs to the government to implement a commercial truck system could be recovered faster than costs of implementing a passenger vehicle system.

- *Reduced reporting burden on trucks.* Mileage fees for commercial trucks could be designed to consolidate existing requirements for interstate commercial trucks to report their miles traveled in most states. Currently, companies registered in states that have established, maintained, or enforced the International Fuel Tax Agreement (IFTA),[59] or in states participating in the International Registration Plan (IRP),[60] are required to track and report their miles traveled in each state to their state of registry.[61] IFTA and IRP provide for the distribution of state fuel and registration taxes among the contiguous 48 U.S. member states and all 10 Canadian provinces based on the number of miles driven by trucks in each state or province. According to Oregon DOT officials, the reporting requirements for these agreements can be administratively time-consuming, particularly for smaller firms that may not have GPS units on board to track their mileage by location. Furthermore, because states have different fuel tax rates, firms may inaccurately report their mileage traveled, particularly in states with higher fuel tax rates than other states. Oregon DOT officials reported that the system tested in that state, or other systems currently available on the commercial market, could be used to assess mileage fees and simplify IFTA and IRP reporting requirements for commercial trucks.

Despite the benefits, opponents of commercial-truck mileage fees in the United States reported that such fees would be burdensome on trucking firms, costly to collect, and difficult to enforce. Specifically, trucking firms would likely incur costs to install mileage fee technology in their vehicles. Trucking industry representatives reported that mileage fees would need to be collected from more than 500,000 firms operating trucks—90 percent of which operate six or fewer trucks—and the cost of compliance could be particularly burdensome to smaller firms. In addition, fees based primarily on self-reported data can be difficult to verify, particularly if the data is not collected electronically. Such fees would also require the state or federal government to monitor and audit trucking firms to ensure the fees reflect the actual mileage logged by each vehicle. Furthermore, trucking industry representatives reported that taxes, such as mileage fees, that require the government to keep proprietary business data without rigorous data protection safeguards in place would be unacceptably intrusive. Commercial truck mileage fee opponents also reported that the fees could result in inequities among trucking firms, with compliant firms paying more than their share of the fees because of any noncompliant competitors.

MILEAGE FEES COULD GENERATE HIGHWAY TRUST FUND REVENUES AND WOULD AFFECT USERS DIFFERENTLY

We modeled the average mileage fee rates that would be needed for passenger vehicles and commercial trucks to meet three illustrative Highway Trust Fund revenue targets ranging from about $34 billion to $78.4 billion per year. To meet these targets, a driver of a passenger vehicle with average fuel efficiency would pay from $108 to $248 per year in mileage fees compared to the $96 they currently pay annually in federal gasoline tax. However, these results do not include the initial startup or ongoing administrative costs of a national mileage fee system in the United States, which are unknown. Mileage fees, like federal fuel taxes, would comprise a small portion of users' overall fuel costs and thus would only marginally increase users' overall transportation costs. For example, drivers of passenger vehicles with average fuel efficiency would pay 7 percent more than their current costs to achieve the highest revenue target of $78.4 billion. Such a change to the fee system, however, would affect users' costs differently based on their vehicle's fuel efficiency, because drivers of less efficient vehicles currently pay more in fuel taxes than drivers of vehicles with greater fuel efficiency. Mileage fees for commercial trucks would also affect users differently because larger trucks tend to be driven more miles and smaller trucks are marginally more fuel efficient. Currently, heavier commercial trucks generally contribute less to the Highway Trust Fund than the costs of their road use. Adjusting the illustrative mileage fee rates to reflect the road damage caused by different vehicles would notably increase commercial truck rates and modestly decrease passenger vehicle rates. However, setting rates that reflect the current costs that different users impose on the system would require up-to-date estimates of vehicles' responsibility for road damage, which are not available.

Using Mileage Fees to Generate Highway Trust Fund Revenues

We modeled three scenarios of average mileage fee rates for passenger vehicles and commercial trucks to replace or augment the Highway Trust Fund revenues currently generated through federal fuel taxes. These scenarios assume that the other federal taxes paid by commercial trucks remain in place and do not include start-up or ongoing administrative costs:[62]

1. *Replace federal gasoline and diesel fuel tax receipts:* In fiscal year 2010, the latest year for which complete data were available, federal gasoline and diesel fuel tax receipts produced nearly $34 billion in revenue. [63]
2. *Meet current spending levels:* Highway Trust Fund user fee revenues have been insufficient to meet authorized spending levels since 2008, and further shortfalls are expected in the years ahead. Using fiscal year 2010 as an example—a year in which Congress augmented the fund with $19.5 billion in general revenues—this scenario models mileage fee rates that could support $53.5 billion in spending.[64] This revenue target is also roughly equivalent to the average annual authorizations for highway and transit programs provided in MAP-21.[65]
3. *Maintain existing conditions and performance levels:* According to the National Surface Transportation Infrastructure Financing Commission, current funding levels will result in further deterioration of the nation's roadways and transit infrastructure. The commission estimated that maintaining the existing infrastructure would require an annual federal contribution of about $78.4 billion for highways and transit. [66]

Mileage fee rates for passenger vehicles and commercial trucks would vary based on the amount of revenue they are set to generate. To meet the federal revenue targets set under the three scenarios we modeled, the average passenger vehicle mileage fee rate would range from less than 1 cent to more than 2 cents per mile and the average commercial truck rate would range from greater than 3 cents to more than 8 cents per mile (see table 2). Converting these mileage fee rates to per-gallon charges to illustrate their approximate relation to current federal fuel tax rates shows that significantly higher fuel taxes would be needed to generate the revenue targets we modeled. For example, converting the mileage fee rates needed to meet current spending levels ($53.5 billion) would translate to a federal gasoline tax of nearly 32 cents per gallon and a federal diesel fuel tax of almost 35 cents per gallon, increases of roughly 72 percent and 43 percent, respectively. For technical details on our modeling, see appendix V.

Table 2. Illustrative Federal Mileage Fee Rates for Three Revenue Scenarios

Revenue scenario	Average passenger vehicle mileage fee (cents per mile)	Comparison with federal gasoline tax[a] (18.4 cents/gal)	Average commercial truck mileage fee (cents per mile)	Comparison with federal diesel fuel tax[b] (24.4 cents/gal)
Replace federal fuel tax receipts ($34 billion)	0.9¢	n/a	3.2¢	n/a
Meet current spending levels ($53.5 billion)	1.5¢	31.6¢ (72% increase)	5.4¢	34.8¢ (43% increase)
Maintain existing conditions and performance ($78.4 billion)	2.2¢	46.6¢ (153% increase)	8.4¢	53.8¢ (120% increase)

Source: GAO analysis.

NOTE: These results do not include the initial start-up or ongoing administrative costs of a national mileage fee system in the United States. Because the results are intended to illustrate the average mileage fees that would be needed to replace federal fuel tax receipts, we assumed that all other federal commercial truck taxes remain in place.

[a]Based on 21.6 mpg average passenger vehicle fuel economy.

[b]Based on 6.4 mpg average commercial truck fuel economy.

These results do not include the initial start-up or ongoing administrative costs of a national mileage fee system in the United States, which are unknown but would result in increased rates. To estimate how such costs would affect mileage fees, we modeled rates that assumed annual fixed costs of 5 percent and 20 percent of current federal fuel tax receipts across all scenarios.[67] We found that the percentage increase in mileage fee rates required to account for costs of implementation is greater with a lower revenue target than with a higher revenue target. For example, assuming 20 percent administrative costs, mileage fee rates for passenger vehicles would need to increase by 27 percent (from 0.9 cents to 1.2 cents per mile) to generate $34 billion and by 13 percent (from 2.2 cents to 2.4 cents per mile) to generate $78.4 billion.

Effects of Mileage Fees on Users

Passenger Vehicles

Mileage fees for passenger vehicles would affect users differently based on their vehicle's fuel efficiency because drivers of less fuel efficient vehicles currently pay more in fuel taxes, as they have to purchase more gasoline to travel the same distance as more efficient vehicles. A system that charges all passenger vehicles the same rate would lead to drivers of more fuel efficient vehicles paying proportionately more in mileage fees than they currently pay in federal fuel taxes. As illustrated in figure 8, a driver of a hybrid with a fuel efficiency of 40 mpg would pay twice as much in mileage fees under the scenario to replace fuel tax receipts and over 4 times as much under the scenario to augment Highway Trust Fund revenues to maintain current conditions and performance. In contrast, a driver of an SUV with a fuel efficiency of 16 mpg would pay less in mileage fees under the scenario to replace federal fuel tax receipts and less than twice as much (88 percent more) under the scenario to augment Highway Trust Fund revenues to maintain current conditions and performance. Although cost increases for passenger vehicle users appear significant under some mileage fee scenarios, increases in their overall transportation costs would be relatively minor because fuel taxes comprise a small portion of the amount that they spend on fuel. For example, under the scenario to maintain current conditions and performance, the owner of a sedan that averages 22 mpg would pay 158 percent more in mileage fees than they pay currently in federal gasoline taxes. However, this represents a relatively small (7 percent) increase in this user's overall transportation costs. Although cost increases under a change to mileage fees are greater for users of more fuel efficient vehicles, a hybrid owner would pay $128 less per month for the combined cost of fuel and mileage fees than the owner of a less efficient SUV in the scenario to meet current conditions and performance, as shown in figure 9. As such, mileage fees would not negate the economic incentives to drive a more fuel efficient vehicle since users would have significantly lower overall fuel costs.

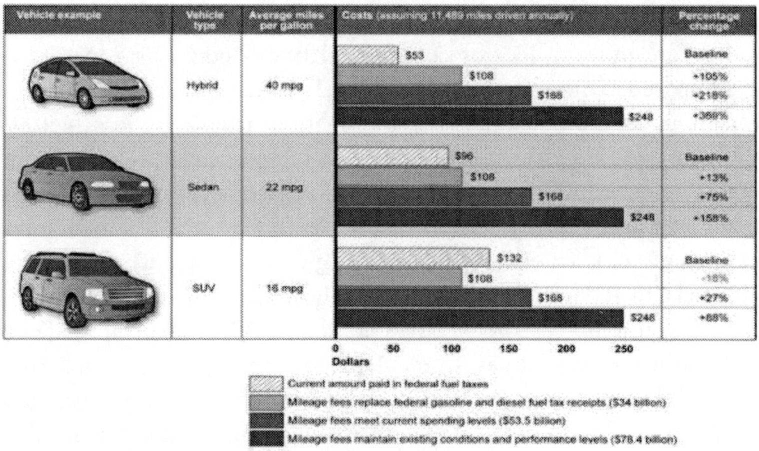

Source: GAO analysis of mileage fee simulation results.

Figure 8. Average Passenger Vehicle Users' Annual Federal Fuel Taxes Paid under Current System Compared with Mileage Fee Scenarios.

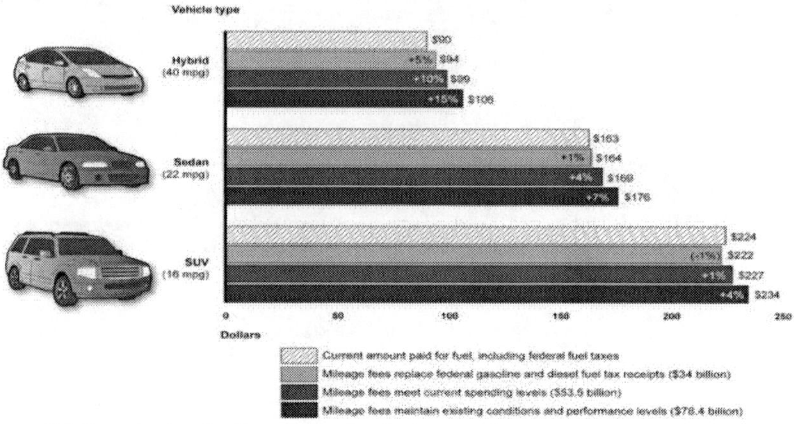

Source: GAO analysis of mileage fee simulation results.

Note: The "current amount paid for fuel" projection assumes a retail gasoline price of $3.75 per gallon, including all state and federal fuel taxes. The three mileage fee scenarios also assume an initial retail gasoline price of $3.75 per gallon, but then replace the 18.4 cent per-gallon federal gasoline tax with mileage fee costs for vehicles of varying fuel efficiency that are based on rates simulated using the average retail gasoline price in 2010. All scenarios assume 11,489 miles driven annually.

Figure 9. Average Passenger Vehicle Users' Monthly Fuel and Tax Costs under Current System and Mileage Fee Scenarios.

Commercial Trucks

A flat-rate mileage fee for commercial trucks could also increase users' costs compared with current diesel fuel taxes but would affect users differently, in part, because larger (combination) trucks are less fuel efficient than smaller (single-unit) trucks. Similar to the passenger vehicle simulation, less fuel efficient combination trucks would pay considerably (35 percent) less under the scenario to replace current federal fuel tax receipts; more fuel efficient single-unit trucks would pay slightly (4 percent) more than they currently pay in federal diesel taxes.[68] In the other two scenarios, single-unit truck costs would increase between 78 percent and 176 percent, while users of combination trucks would pay from 11 percent to 72 percent more. However, just as combination trucks pay far more in diesel fuel taxes in absolute terms, they would pay more in mileage fees than single-unit trucks because they tend to travel significantly more miles annually (see fig. 10).

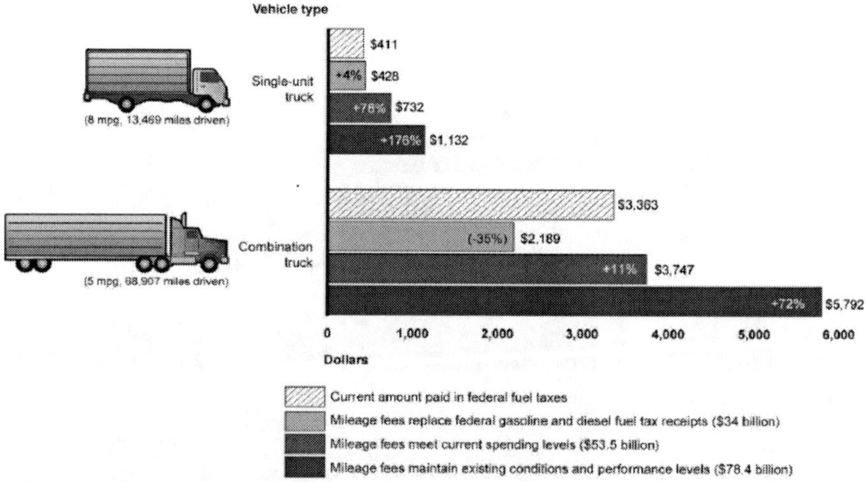

Source: GAO analysis of mileage fee simulation results.

Note: These results assume that the federal heavy vehicle use tax and the federal tire, truck, and trailer excise taxes paid by commercial trucks remain in place.

Figure 10. Average Commercial Truck Users' Annual Federal Fuel Taxes Paid under Current System Compared with Mileage Fee Scenarios.

Similar to passenger vehicles, drivers of commercial trucks would likely pay more in mileage fees than they currently pay in fuel taxes; however, these cost increases represent a smaller portion of their overall costs.[69] For example, while a driver of an average combination truck would pay 72 percent more in mileage fees than he or she currently pays in federal diesel taxes under the scenario to maintain current conditions and performance, this represents a small (4 percent) increase in the driver's monthly costs (see fig. 11).

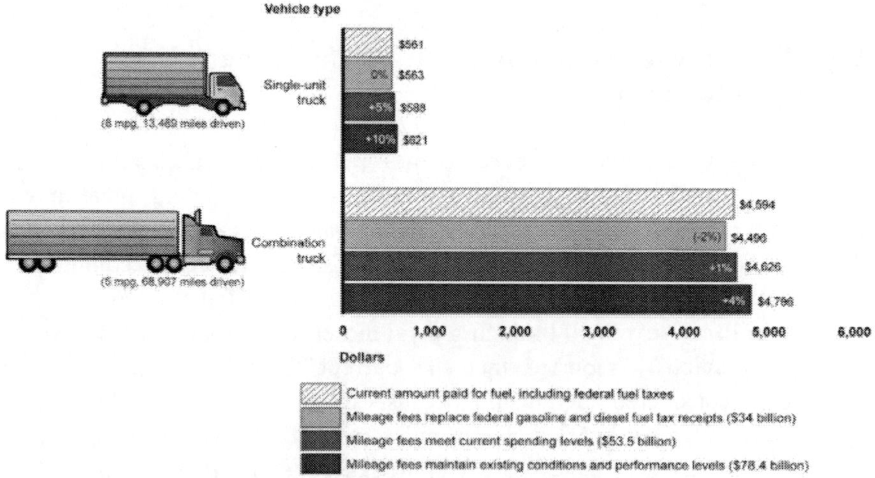

Source: GAO analysis of mileage fee simulation results.

Note: The "current amount paid for fuel" projection assumes a retail diesel fuel price of $4.00 per gallon, including all state and federal diesel fuel taxes. The mileage fee scenarios remove the 24.4- cent federal gasoline tax and replace it with mileage fees. All scenarios assume that other federal taxes paid by commercial trucks remain in place.

Figure 11. Average Commercial Truck Users' Monthly Fuel and Fuel Tax Costs under Current System and Mileage Fee Scenarios.

Although mileage fees could increase costs for owners of commercial truck users, these users would have the ability to defray some of these increases by building the fees into the cost of their services. For example, officials from a German trucking association and a small trucking firm reported that the introduction of HGV tolls has not negatively affected trucking firms because the cost of tolls can be passed along to the consumers of the firm's services. In prior work, we noted that user fees should be charged to the direct user, even if that payer then passes the cost of the fee on to others.[70] We have also noted that, to the extent that costs are not covered by taxes or fees levied on freight providers or consumers, governments would be providing a subsidy to the industry, which is paid by other taxpayers.[71] Although mileage fees could affect the cost of shipping goods, they could, if set at a rate that reflects the cost of road use, promote economic efficiency and minimize the need for the federal government to subsidize transportation funding.

Adjusting Mileage Fees to Account for the Damage Users Impose on Roadways

Although commercial trucks represent 4 percent of the U.S. vehicle fleet, they are responsible for 40 percent of the costs that the federal government spends on highway preservation and maintenance, according to FHWA's *Highway Cost Allocation Study*, published in 2000.[72] However, commercial trucks contributed less than 33 percent of Highway Trust Fund user fee revenues in fiscal year 2010 through federal diesel fuel and other truck taxes.[73] We have previously reported that the current federal highway user fee structure is considered inequitable because it does not effectively capture vehicles' weight per axle or number of miles traveled, the two key components of travel that cause damage.[74] Heavier trucks generally pay less than their share of damage costs because the current federal tax structure does not fully account for the increased road wear caused by heavy trucks based on their miles traveled and weight.[75] The federal diesel fuel tax—the primary source of commercial trucks' Highway Trust Fund contributions—charges all trucks the same rate of 24.4 cents per gallon, regardless of vehicle weight or number of axles. Additionally, the federal heavy vehicle use tax is capped at $550 for trucks weighing more than 75,000 pounds, meaning that all trucks above this weight pay the same fee despite the increased damage costs the heavier trucks impose.[76] In 1994, we recommended that Congress consider a national weight-distance user fee to increase equity and promote a more efficient use of the nation's highways.[77] Based on FHWA's 2000 estimates, commercial trucks would need to contribute 40 percent of the revenues deposited in the Highway Trust Fund—as opposed to less than 33 percent in fiscal year 2010—to cover the costs they impose on the nation's roadways, leaving passenger vehicles to contribute about 60 percent.

Adjusting the illustrative mileage fee rates so that users' contributions to the Highway Trust Fund reflect FHWA's 2000 cost estimates would result in substantial rate increases for commercial trucks and modest decreases for passenger vehicles. As shown in figure 12, adjusted commercial truck rates would range from approximately 4 cents to 11 cents per mile, compared with the original range of about 3 cents to 8 cents per mile— increases of approximately 30 percent in each scenario. 78 These rate increases are large because the increase in Highway Trust Fund contributions would be spread across 10 million commercial trucks, or 4 percent of the U.S. vehicle fleet. Conversely, the adjusted passenger vehicle mileage fee rates would decrease modestly because the savings would be spread across the U.S. fleet of 230 million passenger vehicles. The adjusted rates account for the $2.8 billion paid in fiscal year 2010 by commercial trucks in federal truck taxes, which would still be needed to ensure that trucks contribute 40 percent of Highway Trust Fund revenues. Mileage fees would continue to represent a small portion of users' overall transportation costs.

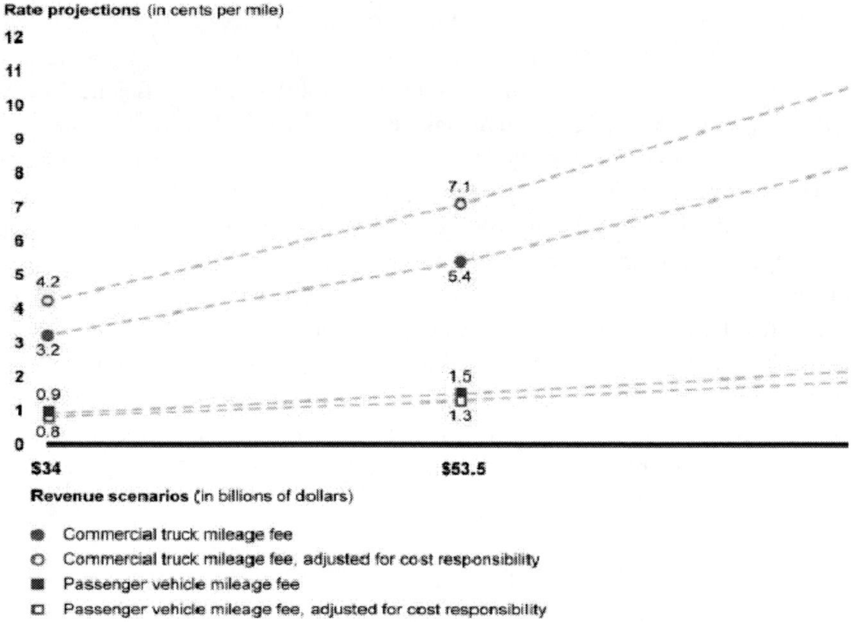

Source: GAO mileage fee simulation results.

Note: Our simulation produced mileage fee rates specific to three revenue targets and did not produce rates for any revenue amounts between the targets. As such, the lines in the figure above are meant to illustrate the degree of change between the rates in each of the three scenarios. The adjusted rates account for the $2.8 billion paid in fiscal year 2010 by commercial trucks in federal truck taxes, which would still be needed to ensure that trucks contribute 40 percent of Highway Trust Fund revenues.

Figure 12. Illustrative Federal Mileage Fee Rates for Three Revenue Scenarios, Adjusted for Cost Responsibility.

Establishing user fee rates that reflect the current costs that different users impose on the system would require up-to-date estimates of vehicles' responsibility for road damage. FHWA conducted its most recent full revision of its *Highway Cost Allocation Study* in 1997 in response to a GAO recommendation,[79] and it issued an addendum to the study in 2000 with updated results. However, FHWA officials reported that it is likely that the estimated share of commercial trucks' responsibility for road damage has changed over the last 12 years because of increases in miles traveled by heavy trucks relative to smaller trucks and cars, shifts in federal spending, and other factors. In addition, trucking industry representatives told us that an updated *Highway Cost Allocation Study* would better reflect current conditions and show that commercial truck revenues to the Highway Trust Fund have increased in recent years relative to passenger vehicle revenues. This is because of improvements in passenger vehicles' fuel efficiency, while commercial truck fuel efficiency has remained flat. Moreover, some states have been granted waivers to permit larger and heavier commercial trucks to travel on Interstate highways, which could increase the amount of damage caused by the vehicles. Prior GAO work has shown that designing a user fee system with unreliable cost information can skew fee-setting decisions.[80] Furthermore, up-to-date information on the on the costs imposed by different users in relation to the revenues they contribute to the Highway Trust Fund could be used to evaluate whether the rates paid by users through federal motor fuel or truck taxes are sufficient. FHWA officials said they do not periodically update the *Highway Cost Allocation Study* and have no plans to resume work on a new study at this time.

STATES RECOGNIZE THAT ALTERNATIVE TRANSPORTATION REVENUES ARE NEEDED, AND MANY WOULD SUPPORT FEDERAL ACTIONS TO EVALUATE MILEAGE FEE SYSTEMS

Fifty of the 51 state DOTs we surveyed agreed that it is important that an alternative federal funding mechanism be identified in the next 10 years in order to meet surface transportation revenue needs.[81] However, only 8 of the 51 states reported that they are likely to introduce some type of mileage fee program in the next 10 years. Of these, 7 reported that they are likely to introduce mileage fees for electric vehicles in the next decade and fewer reported that they are likely to introduce programs for passenger vehicles (4) or commercial trucks (3). Although such efforts could produce transportation revenues for these states, they would not generate federal revenues for the Highway Trust Fund. However, more than half of all states responded that they would support federally-led field tests to evaluate mileage fees. There are currently no federal pilot programs or other efforts to evaluate the viability and cost of implementing such a system in the United States.

The majority of state DOTs responded that they would support federal field tests of mileage fees for electric vehicles and commercial trucks. Proponents of mileage fees have suggested that charging drivers of these vehicles for their road use could improve the equity and sustainability of Highway Trust Fund revenues:

- *Electric vehicles.* Two-thirds of state DOTs (34 of 51) reported they would support federally-led field tests of mileage fees for electric vehicles; none reported that they would be opposed to field tests for these vehicles. Electric vehicles can help achieve important policy goals to reduce emissions and limit U.S. dependence on foreign oil; however, these vehicles do not use gasoline or diesel fuel or contribute to the Highway Trust Fund through any other federal fees or taxes. The Department of Energy estimated that there were fewer than 60,000 electric vehicles in the United States in 2010. Although charging electric vehicle mileage fees would produce minimal revenues in the immediate future, the Department of Energy estimates that more than 1.2 million electric vehicles may be produced by 2015.[82] In focus groups conducted by the Texas Transportation Institute, participants recognized that a potentially large segment of the future vehicle fleet may pay nothing aside from state vehicle registration and title fees for their road use and strongly preferred mileage fees for this class of vehicles.[83] An Oregon DOT official reported that the state is planning to introduce legislation to pilot user fees for electric vehicles in 2013.
- *Commercial trucks.* Almost 60 percent of state DOTs (30 of 51) reported that they would support federally-led field tests of mileage fees for commercial trucks. To date, very few states have evaluated mileage fees for commercial trucks. In a March 2011 report to Congress, FHWA reported that additional technical research would be required to assess methods to charge heavy trucks based on their infrastructure wear. According to FHWA, charging trucks based on

weight and distance traveled has been recognized as the best way to reflect the road damages caused by different types of trucks.[84] Although the trucking industry has resisted weight-distance taxes, FHWA reported that many of their objections, including administrative burden, can be overcome if linked with a mileage fee system. However, other technological issues would need to be resolved, such as how to measure the weight of combination vehicles that pull different types and numbers of trailers. FHWA reported that the federal government has a leadership role to play to prevent different systems from being established across states that later could not be reconciled into a national mileage fee. Such a role could include field tests to determine how and whether a mileage fee system should be implemented, including at least one trial for commercial trucks, and the evaluation of the costs and benefits of various approaches.

While states broadly supported federal actions to explore mileage fees, more states placed a high priority on increasing federal fuel taxes to meet surface transportation revenue needs.

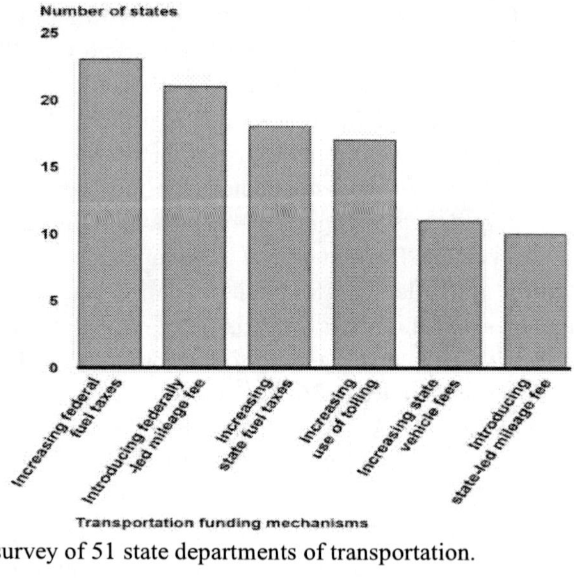

Source: GAO survey of 51 state departments of transportation.

Figure 13. Number of State DOTs Reporting High Priority on Various Transportation Funding Mechanisms, 2012.

As shown in figure 13, 23 states reported that they place a high level of priority on increasing federal fuel taxes; 21 states place a high priority on introducing a federally-led mileage fee. Prior GAO work has shown that a higher gasoline tax could encourage drivers to reduce fuel consumption by driving less, reduce the nation's dependence on oil, relieve highway congestion, and decrease emissions of gases that pollute the air.[85] However, as vehicles become more fuel efficient and increasingly run on alternative fuels, fuel taxes may not be a long-term source of transportation funding.

CONCLUSION

The nation's surface transportation system is under growing strain, and the costs of repairs and upgrades to meet current and future demands are estimated in the hundreds of billions of dollars.

The ongoing erosion of federal fuel tax revenues will worsen in the years ahead with the introduction and adoption of more fuel efficient and alternative fuel vehicles.

Users of the surface transportation system pay less for their use of roadways in federal fuel taxes and user fees than the federal government now spends to maintain and improve the system. The use of general revenues to cover Highway Trust Fund spending breaks the link between highway taxes paid and benefits received by users and may not be a sustainable strategy, given competing demands for federal funds and the federal government's growing fiscal challenge.

Technologies are currently available to gather mileage data and charge users mileage fees, should Congress wish to explore such fees as a means to help address shortfalls in surface transportation funding. However, the perception that these technologies will be used to track privately-owned vehicles and infringe upon individual privacy currently appears to be an insurmountable challenge.

Because the public perception of privacy risks would be particularly acute in mileage fee systems that mandate the use of GPS technologies, the widespread implementation of such a system to cover all U.S. passenger vehicles appears unlikely at this time. Although technology evolves rapidly and public perception can change over time, it may be impractical for the federal government to pursue mileage fees for all vehicles through a system that collects and reports information on people's movements for the purpose of assessing taxes.

Nonetheless, there may be opportunities to evaluate mileage fees for certain types of vehicles that could improve the equity of highway funding and begin to address Highway Trust Fund shortfalls. The current U.S. fuel tax system does not reflect the costs of road use and the anticipated future changes in the U.S. vehicle fleet. Germany and New Zealand have demonstrated that variable rate, distance-based user fees for commercial trucks can generate substantial revenues linked to road damage costs and help reduce emissions while posing fewer privacy concerns than passenger vehicle systems. In the United States, commercial trucks' contributions to the Highway Trust Fund through federal diesel fuel and related truck taxes do not reflect the relationship between truck weight and road damage. In addition, although few alternative fuel vehicles are on the roads today, their numbers are expected to increase significantly in the coming years, which could provide environmental and foreign policy benefits to the country by reducing U.S. dependence on foreign oil.

However, drivers of these vehicles do not currently contribute to the Highway Trust Fund through user fees or fuel taxes for their road use. State DOTs reported broad support for federal initiatives to evaluate federal mileage fees, including federally-led field tests for electric vehicles and commercial trucks. State pilot programs have shown the feasibility of assessing mileage fees, but the programs have not shed light on how mileage fees could be implemented in an integrated fashion across states to provide revenues to the federal Highway Trust Fund. Furthermore, state efforts to evaluate mileage fees for commercial trucks have yet to determine the best approaches to address trucking industry concerns regarding administrative burden and compliance costs.

A federal pilot program provides the opportunity to assess balancing those concerns with the goal of ensuring that federal fees cover the costs of users' road use. In the absence of any current federal pilot programs or efforts to evaluate (1) options to more accurately charge commercial trucks and electric vehicles for their road use and (2) the cost to launch and administer such systems, Congress lacks critical information to assess whether mileage fees for these vehicles could be a viable and cost-effective tool to begin to address federal surface transportation funding challenges. Congress and the Administration have yet to develop a long-term plan for funding surface transportation; ultimately, increased surface transportation revenues, reduced transportation spending levels, or both will be needed to bring revenues and spending into balance.

Consideration of new revenues—whether through mileage fees, fuel taxes, or other fees—would benefit from accurate and up-to-date information on the damage different vehicles impose on the roadways and whether all vehicles are currently paying their fair share.

The most recent FHWA estimates from 2000 suggest that many commercial trucks underpay, but these estimates may not reflect current conditions. In the absence of periodically updated estimates from FHWA, Congress lacks the information necessary to determine whether revenues collected from different users are sufficient to cover the costs of their road use. Should Congress wish to explore mileage fees, or other sources of revenue, updated information would allow it to consider the costs imposed by different users in setting appropriate rates.

MATTERS FOR CONGRESSIONAL CONSIDERATION

Should Congress wish to explore mileage fees as a mechanism for funding surface transportation, it should consider establishing a pilot program to evaluate the viability, costs, and benefits of mileage fee systems for:

- commercial trucks—to ensure that fees paid by the owners of these vehicles cover the costs of their use of the nation's roads and bridges, and
- electric vehicles—to develop a mechanism through which the owners of these vehicles can contribute to the Highway Trust Fund for their use of the nation's roadways.

RECOMMENDATION FOR EXECUTIVE ACTION

To ensure that up-to-date data are available on the road damages imposed by all vehicles types compared with the revenues each contributes to the Highway Trust Fund, we recommend that the Secretary of Transportation direct the FHWA Administrator to revise and publish the agency's *Highway Cost Allocation Study* and update it periodically as warranted.

AGENCY COMMENTS

We submitted a draft of this report to the Department of Transportation for review and comment. The department did not take a position on our recommendation to revise and publish the *Highway Cost Allocation Study* and update it periodically as warranted. The department provided technical comments via email which we incorporated as appropriate.

Susan Fleming
Director,
Physical Infrastructure Issues

APPENDIX I: OBJECTIVES, SCOPE, AND METHODOLOGY

To assess the benefits achieved and challenges faced in mileage-based user fee (mileage fee) initiatives in the United States and selected other nations, we reviewed academic literature to identify the range of mileage fee pilot projects conducted in the United States and distance-based user fee programs introduced internationally. We interviewed the lead researchers and reviewed the reported results of all U.S.-based mileage fee pilot projects conducted as of June 2012 including: (1) a National Evaluation of Mileage-based Road User Charges, conducted in 12 states from 2008 to 2010 by the University of Iowa, as authorized by Congress through the Safe, Accountable, Flexible, Efficient Transportation Equity Act: A Legacy for Users (SAFETEA-LU)[1]; (2) the Oregon Road User Fee Pilot Program Oregon, conducted by the Oregon Department of Transportation (DOT) from 2006 to 2007; and (3) the Traffic Choices Study, conducted by the Puget Sound Regional Council in Washington state from 2005 to 2006. We also reviewed preliminary results from two on-going pilot projects conducted by the Minnesota DOT and Nevada DOT and interviewed transportation officials involved in these pilots.[2]

Through our literature review, we identified distance-based user fee initiatives in Germany, New Zealand, the Czech Republic, Switzerland, Austria, Slovakia, and the Netherlands. We selected three international programs for our review: (1) the German Heavy Goods Vehicle (HGV) Tolling system, which charges commercial trucks over 26,000 pounds a distance-based fee for travel on the national motorway, or autobahn; (2) the

New Zealand Road User Charge (RUC) system, which charges all commercial trucks over 7,700 pounds and all diesel-fueled passenger vehicles for distances traveled on that nation's public roadways; and (3) a proposed user fee initiative in the Netherlands that was suspended before it was implemented that would have charged all commercial trucks and passenger vehicles distance-based fees. We selected these programs using criteria that included the primary objectives of each program as described in the literature, the types of technology used, the classes of vehicles covered, and the administration of the program, such as the use of the private sector to implement the system. We included the suspended initiative in the Netherlands in our review to assess the challenges of implementing a national user fee program. Because we used a nongeneralizable sample to select the programs to review, our findings are not representative of all countries with distance-based user fee programs. We reviewed data on the benefits achieved by these programs, including revenues raised, and the challenges faced in the programs, including their costs of implementation. We conducted a site visit in Berlin, Germany and interviewed government officials and private sector representatives from New Zealand and the Netherlands via teleconference.

To determine the mileage fee rates necessary to replace and supplement current fuel tax revenues deposited in the Highway Trust Fund and the effect these fees would have on users' costs, we conducted an economic simulation to produce illustrative rates for passenger vehicles and commercial trucks, an approach that is commonly used in relevant existing studies. To estimate the mileage fee rates, we focused the hypothetical scenarios on three target levels of revenue, namely: (1) scenario 1, in which mileage fee revenues would simply replace federal gasoline and diesel fuel tax receipts deposited in the Highway Trust Fund ($34 billion in fiscal year 2010); (2) scenario 2, in which mileage fees would generate revenues sufficient to meet current spending levels ($53.5 billion in fiscal year 2010); and (3) scenario 3, in which mileage fees would generate the $78.4 billion necessary to maintain existing level of conditions and performance, assuming a 2 percent inflation rate going forward. In all three scenarios, we assume that the federal heavy vehicle use tax and the federal tire, truck and trailer excise taxes paid by commercial trucks ($2.8 billion in fiscal year 2010) would remain in place.

We also simulated mileage fee rates that assumed ongoing administrative costs of 5 percent of revenues and 20 percent of revenues. To do so, we increased the revenue target in scenario 1 by 5 percent ($1.7 billion) and 20 percent ($6.8 billion) and simulated mileage fee rates using the same process described in this appendix and in appendix V. We then applied the same cost assumptions—$1.7 billion and $6.8 billion—to the revenue targets in scenarios 2 and 3 so that the ongoing administrative costs would be fixed in all scenarios and simulated mileage fee rates using the same process described above.

The target revenues in scenario 1 were based on fiscal year 2010 federal gasoline and diesel fuel tax receipts as reported in the Federal Highway Administration's (FHWA) *Highway Statistics 2010* report, the most current available data at the time of our review.[3] In 2010, Congress directed $19.5 billion from the General Fund to the Highway Trust Fund and we used this to compute the $53.5 billion revenue target in scenario 2. For scenario 3, we relied on a forecast estimate by the National Surface Transportation Infrastructure Financing Commission (NSTIFC) report that $78.4 billion in federal revenues (in 2008 dollars) would be required to maintain the existing conditions and performance of the nation's highway and transit system.[4] Using NSTIFC's assumption of a constant inflation rate of 2 percent, we estimated it would have required approximately $81.2 billion to maintain the existing conditions and performance of the nation's highway and transit system in fiscal year 2010.[5] To be consistent with the first two scenarios, in which we excluded the $2.8 billion of federal heavy vehicle use tax and the federal tire, and truck and trailer excise taxes paid by commercial trucks in fiscal year 2010 from the revenue targets, we subtracted $2.8 billion from the forecast revenue ($81.2 billion) to obtain a final target revenue of $78.4 billion.

To determine the respective shares of mileage fee revenues that passenger vehicles and commercial trucks would need to contribute, we looked to the amount each group contributed in federal fuel taxes in fiscal year 2010. First, we added the $2.8 billion that commercial trucks paid in the federal heavy vehicle use tax and the federal tire, and truck and trailer excise taxes in fiscal year 2010 to the federal diesel fuel tax revenues in the revenue targets to reflect the total amount of federal taxes these vehicles pay for their road use. Doing so temporarily increased the revenue targets to $36.7 billion in scenario 1, $56.2 billion in scenario 2, and $81.2 billion in scenario 3. Next, because the $36.7 billion deposited in the Highway Trust Fund in 2010 was comprised of $24.8 billion (67.6 percent) of federal gasoline tax revenues and $11.9 billion (32.4 percent) of federal diesel fuel tax revenues and other commercial truck taxes, we applied a ratio of 67.6 percent (passenger vehicles) and 32.4 percent (commercial trucks) to the overall revenue targets in scenarios 2 and 3. We then subtracted $2.8 billion from the resultant commercial truck figures to arrive at their revenue contribution, producing the following total revenue targets: $34 billion in scenario 1, $53.5 billion in scenario 2, and $78.4 billion in scenario 3. This final step ensured that the $2.8 billion in non-diesel tax revenues that commercial trucks contribute to the Highway Trust Fund are not considered in our simulation. Our results are intended to illustrate the average mileage fees that would be needed to replace federal fuel tax receipts and we assume that all other federal commercial truck taxes remain in place.

In theory, it would be possible to set mileage user fee rates to vary within different classes of vehicles and by location, time of day, or type of road. Targeted, variable mileage user fees could provide financial incentives for users to change their existing behavior to reduce what economists call externality costs, such as traffic congestion and vehicle emissions. If designed appropriately, finely targeted mileage fees could also maintain or even augment the incentive for motorists to buy more fuel efficient vehicles if less efficient vehicles were charged greater rates. However, as a result of our decision to consider only two general types of vehicles, our simulation model only generates two flat mileage fee rates that represent an average rate across each of the two vehicle types in each revenue scenario: one for passenger vehicles and one for commercial trucks. To adjust our simulation's illustrative mileage fees for passenger vehicles and commercial trucks to reflect the respective road damage costs they impose, we used the cost responsibility ratio from the 2000 update to the *1997 Highway Cost Allocation Study* compiled by FHWA to adjust each scenario's revenue targets for passenger vehicles and commercial trucks.[6] FHWA's cost responsibility ratio measures the responsibility of different vehicle classes for highway program costs paid from the Highway Trust Fund to the federal user fees paid by the different vehicle classes. To adjust the revenue targets for passenger vehicles and commercial trucks in each of the 3 scenarios, we first added back the federal tire tax, truck and trailer excise tax, and heavy vehicle use tax ($2.8 billion in total) to the commercial truck revenue targets to reflect the amount of federal taxes that these vehicles pay, in total, for their road use. This increased the overall revenue targets to $36.7 billion in scenario 1, $56.2 billion in scenario 2, and $81.2 billion in scenario 3. We then applied FHWA's cost responsibility ratios of 59.7 percent (passenger vehicles) and 40.3 percent (commercial trucks) to the each scenario's overall revenue target to obtain the total amount of Highway Trust Fund revenues that would be required from each group. This ensured that the amount commercial trucks paid in federal taxes in addition to federal diesel fuel tax revenues in fiscal year 2010 were considered as part of their overall contribution. We then subtracted $2.8 billion from the resultant figures for trucks to identify their revenue contribution through only federal diesel fuel tax and arrive at the adjusted revenue targets. Adding $2.8 billion to any of the three commercial truck revenue targets would then result in this group's total contribution equaling 40.3 percent of each scenario's overall revenue target; each corresponding passenger vehicle revenue target would comprise the other 59.7 percent of each scenario's overall target. We recognize that the 40.3 percent cost responsibility ratio is an average across all commercial truck configurations and weights.

Table 3. Values of Key Parameters Used in Simulation

Parameters	Scenario	Passenger vehicles	Commercial Trucks	Total
Average fuel efficiency[a] (miles per gallon)	All	21.6	6.4	—
Fuel tax[a] (cents per gallon)	All	18.4	24.4	—
Vehicle miles traveled[a] (VMT) (miles)	All	2,647,659,000,000	286,585,000,000	2,934,244,000,000
Average fuel price[b] (dollars per gallon)	All	2.84	2.99	—
VMT elasticity to fuel price[c]	All	-0.40	-0.24	—
Revenue targets[d]	Scenario 1	$24,836,919,000	$9,135,819,000[e]	$33,972,738,000
	Scenario 2	$38,019,301,988	$15,453,436,012	$53,472,738,000
	Scenario 3	$54,859,805,044	$23,524,195,956	$78,384,001,000
Revenue targets[f] (adjusted for cost responsibility)	Scenario 1	$21,933,742,389	$12,038,995,611	$33,972,738,000
	Scenario 2	$33,575,242,389	$19,897,495,611	$53,472,738,000
	Scenario 3	$48,447,266,400	$29,936,734,600	$78,384,001,000

Source: GAO analysis of FHWA data.

[a]FHWA, *Highway Statistics 2010*, Table VM1, (Washington, D.C., 2012). We did not include VMT for motorcycles or buses, which are not included in the FHWA definitions of light duty vehicles, single-unit trucks, or combination trucks. In addition, buses are exempt from most highway user taxes. [b]U.S. Energy Information Agency data (2010).

[c]Parry, Ian and Kenneth Small, Does Britain or the United States Have the Right Gasoline Tax?, *American Economic Review*, Vol. 95. No.4, page 1283 (2005). Elasticity measures the extent VMT may change as a percentage in response to a one-percent increase in fuel price. For example, a VMT elasticity to fuel price of -0.40 means that a one percent increase in fuel price will induce a decrease of 0.40 in VMT.

[d]FHWA, *Highway Statistics 2010*, Table FE10, (Washington, D.C., 2012).

[e]According to FHWA, illegal evasion of the diesel fuel tax might have resulted in sizeable loss of revenue from $1 billion to as much as 25 percent of total revenue.

[f]Cost responsibility ratio from FHWA, *Addendum to the 1997 Highway Cost Allocation Study,* Table 4, (Washington, D.C., May 2000).

To more accurately charge trucks fees that reflect the damages they impose, variable mileage fees would need to be applied based on vehicle weight and number of axles. To ensure that the key assumptions we made were reasonable, we calibrated the simulation using existing literature and data from government or other credible sources. We used data on vehicle miles traveled (VMT), fuel consumption, and federal gasoline and diesel tax receipts from FHWA *Highway Statistics 2010* report, and utilized 2010 data on the average retail gasoline and diesel fuel price compiled by the U.S. Department of Energy's Energy Information Agency. Table 3 provides the key parameters used in our simulation as well as all passenger vehicle and commercial truck revenue targets, including those adjusted to reflect cost responsibility.

To conduct our analysis, we made a few simplifying assumptions:

1. All the light duty vehicles reported in FHWA's *Highway Statistics 2010* are passenger vehicles to facilitate comparison with FHWA's *Highway Cost Allocation Study*;[7]
2. All federal gasoline tax receipts are generated by passenger vehicles;
3. All federal diesel fuel tax receipts are generated by commercial trucks;[8]
4. If implemented, all users would pay the mileage fees (no evasion).

We anticipated the effects of these assumptions on revenues and mileage fee rates to be small. For example, according to the Environmental Protection Agency, only 0.74 percent of all model year 2010 light duty vehicles in the United States ran on diesel fuel and, as result, the amount of federal diesel fuel tax receipts contributed to the Highway Trust Fund by passenger vehicles is minimal.[9]

To ensure that the simulation analysis generated reasonable results, we conducted a sensitivity analysis to examine how the mileage fees would have changed with respect to three key parameters: (1) fuel price, (2) VMT elasticity to fuel price, and (3) average vehicle fuel efficiency. The results of the analysis do not indicate any unexpected or significant changes in mileage fees within the examined range of parameter values.

In addition, we compared our analysis results with findings from other existing studies of mileage fees and found our results to be consistent with the existing literature.[10] Please refer to appendix V for a detailed description of the simulation analysis and its limitations.

To examine the effect that the illustrative mileage fee rates produced by our simulation would have on users' costs, we compared the annual costs that average drivers of passenger vehicles and commercial trucks currently pay in federal fuel taxes with the costs each would expect to pay in each of the three revenue scenarios. For passenger vehicles, we first divided the federal gasoline tax of 18.4 cents per gallon by varying levels of vehicle fuel economy to establish different users' effective mileage fee rates under the current system. We then multiplied each of these rates— as well as the illustrative mileage fee rates from our simulation—by the average number of miles driven annually by a passenger vehicle (11,489) to determine the average amount different users would pay in federal fuel taxes under the current system or in mileage fees in the three revenue scenarios. In addition, we also divided the average number of miles driven annually by varying levels of vehicle fuel economy to determine how many gallons of gasoline different users would expect to purchase each year. We then multiplied this number by $3.57 to calculate users' average fuel costs without federal taxes or mileage fees, having assumed a retail price of gasoline of $3.75 and subtracted the federal gasoline tax of 18.4 cents. To examine the effect of mileage fees for drivers of commercial trucks, we repeated the same steps using the federal diesel tax rate of 24.4 cents per gallon; the commercial truck mileage fee rates produced by our simulation; estimates for fuel economy and number of miles driven annually for single-unit trucks (8 mpg; 13,469 miles per year) and combination trucks (5 mpg; 68,907 miles per year) from the National Academy of Sciences[11] and FHWA's *Highway Statistics 2010* report; and an assumed retail diesel fuel price of $4.00 per gallon.

To identify states' views on addressing future revenue demands using mileage fees, we surveyed the DOTs in all 50 states and the District of Columbia and received a 100 percent response rate. We pretested the survey with officials from four state departments of transportation and obtained feedback from FHWA officials to ensure that questions were clear, unbiased, comprehensive, and that terminology was used correctly. We made changes to the content of the questions in response to the pretests and FHWA review. Because we administered the survey to the complete universe of potential respondents, there are no sampling errors. However, the practical difficulties of conducting any survey may introduce errors, commonly referred to as non-sampling errors. For example, difficulties in how a particular question is interpreted, in the sources of information that are available to respondents, or in how the data is entered into a database or analyzed can introduce unwanted variability into the survey results. To minimize these types of errors, we employed recognized survey design practices in developing the questionnaire during pretesting and throughout the collection, processing, and analysis of the survey data. For example, in reviewing the survey data, we performed checks to identify missing or ambiguous responses and we addressed these errors by contacting the officials involved to clarify their responses. We conducted our survey from March 22 through April 24, 2012. See Appendix VI for a summary of survey results.

We conducted this performance audit from December 2011 through December 2012 in accordance with generally accepted government auditing standards. Those standards require that we plan and perform the audit to obtain sufficient, appropriate evidence to provide a reasonable basis for our findings and conclusions based on our audit objectives. We believe that the evidence obtained provides a reasonable basis for our findings and conclusions based on our audit objectives.

APPENDIX II: U.S.-BASED PILOT PROJECTS ON MILEAGE-BASED USER FEES

Table 4. Summary of Mileage-based User Fee Pilot Projects in the United States, as of June 2012

Pilot name and location(s)	Type of vehicles	Summary of methodology
National Evaluation of a Mileage-based Road User Charge (Iowa study) Field tests in 12 states: California, Florida, Idaho, Illinois, Iowa, Kansas, Maine, Maryland, Montana, New Mexico, North Carolina, and Texas	Passenger vehicles	Researchers from the University of Iowa Public Policy Center tested a prototype mileage fee system in the vehicles of approximately 2600 volunteer participants. The tests were conducted in two 10-month test phases from October 2008 through June 2010. Participants were surveyed before, during, and after the tests to gauge their acceptance of the fees. The system used a Global Positioning System (GPS)-based, on-board unit that computed hypothetical mileage fees for federal, state, and, whereapplicable, local jurisdictions and periodically uploaded charges over a cellular communications link to a central office billing center. The billing center created monthly mileage charge invoices that were sent to participants. Officials reported that the system did not retain or transmit any GPS coordinates or other specific information regarding vehicle location or routes travelled. According to the final report, researchers found that mileage-based road-user charging is technically feasible using currently mature technologies; however, the installation of charging equipment into existing vehicles may pose a significant challenge.

Table 4. (Continued)

Pilot name and location(s)	Type of vehicles	Summary of methodology
Oregon Road User Fee Pilot Program Portland, Oregon	Passenger vehicles	In April 2006, the Oregon DOT conducted a 12-month field test with 285 volunteer vehicles, using a prototype, pay-at-the-pump mileage fee system. This study equipped vehicles with a GPS receiver to determine hypothetical mileage charges based on several zones (in-state, out-of-state, and within the Portland metropolitan area) and a wireless transmitter used to transmit mileage information to the fueling station's point-of-sale system at a participating gas station equipped with wireless receivers. When participants filled their tanks with fuel, the fueling station's point of sale system sent total mileage by zone to Oregon State University for processing. Participants' fuel receipts showed the difference in a hypothetical payment in a mileage fee system. According to the final report, the zone pricing strategy applied in the pilot program produced a 22 percent decline in driving during peak periods.
Puget Sound Regional Council (PSRC), Traffic Choices Study Seattle, Washington	Passenger vehicles	From July 2005 through February 2006, PSRC installed GPS tolling meters in the vehicles of 275 volunteer households to observe driving patterns before and after hypothetical tolls were charged for the use of major freeways and arterials in the Seattle metropolitan area. The on-board unit displayed the specific toll rates applicable based on the road type and time-of-day. The meter stored latitude and longitude coordinates and toll data, and periodically transmitted it to a central office using cellular communications. The central office deducted hypothetical chargesfrom a user account populated with funds to cover the costs of travel based on their typical driving habits. If participants varied their travel to avoid congestion charges, they would be charged a lower rate for their travel and could keep the balance of their accounts after the testing period. According to the study's summary report, participants made small-scale adjustments in travel that, in aggregate, would have a major effect on transportation system performance. The study suggests there is an opportunity to significantly reduce traffic congestion and raise revenues for investment using road charging.
Minnesota Department of Transportation Road Use Test St. Paul, Minnesota	Passenger vehicles	At the time of our review, Minnesota DOT was in the process of testing a mileage fee system that uses commercially-available, smart phones with GPS receivers and mileage-metering applications developed by Minnesota DOT. Minnesota DOT installed the smart-phones in a total of 500 vehicles in three 6-month test phases.

Pilot name and location(s)	Type of vehicles	Summary of methodology
		Drivers received funds from Minnesota DOT near the beginning of their test period based upon their typical driving habits and used those funds to pay monthly mileage fee bills sent by Minnesota DOT. The project does not collect actual revenues from participants as they do not pay any mileage fees that may be incurred in excess of the amount in their Minnesota DOT accounts. If participants drive fewer miles than expected, they are allowed to keep anyremaining funds. Rates per mile vary based on location from: (a) 3-cents per mile inside the Minneapolis-St. Paul metropolitan area during peak, time-of-day travel periods; (b) 1-cent per mile during off-peak travel in the Minneapolis-St. Paul metropolitan area and during any time outside the region but in-state; and (c) no charges for out-of-state travel. Participants can turn off the smart phone at any time to not have their mileage metered. However, Minnesota DOT reads participant odometers at the mid-point and at the end of the study, and participants are charged 3 cents per mile for any mileage on the odometer that was not captured on the smart phone system. Minnesota DOT plans to complete the tests by the end of 2012 and issue a final report of the results soon after.
Nevada Department of Transportation VMT Fee Study Las Vegas, Nevada	Passenger vehicles	At the time of our review, Nevada DOT was conducting a mini-test with 35 vehicles using a pay-at-the pump system to prepare for a larger pilot to be conducted in 2013. The system to be tested does not use GPS. Instead, awireless transponder is connected to vehicle's on-board diagnostics unit to gather mileage data and transmit it to a wireless receiver at the fuel pump. The information is sent via an Internet connection to a central officecomputer, which calculates mileage fee based on the total, undifferentiated miles traveled. The central office sends back the hypothetical mileage fee which is included in the fuel receipt.
Oregon Truck Road Use Electronics (TRUE) Project Oregon	Commercial trucks	In 2010, Oregon DOT conducted field tests of an electronic payment system for Oregon's existing weight-mile tax. Twenty-five participating trucks from three Oregon-based trucking firms were equipped with a smart phone and weight-distance tax application. Drivers input the vehicle weight and axle configuration at the onset of the trip into the smart phone, which recorded the vehicle's mileage data. The device sent the information via cellular connection to Oregon DOT, which used the data topopulate a website from which a participating motor carrier could view their own data only and create and transmit their monthly highway use tax bills to the state of Oregon.

Source: GAO summary of pilot project documentation.

APPENDIX III: USER FEE DESIGN CRITERIA

Prior GAO work has found that the design of user fee programs can be evaluated based on several related criteria: efficiency, equity, revenue adequacy, and administrative burden.' These criteria interact and are often in conflict with each other; as such, there are trade-offs to consider among the criteria when designing a fee (See Table 5.)

Table 5. GAO Criteria for User Fee Design

Criteria	Description
Efficiency	By requiring identifiable beneficiaries to pay for the costs of services, user fees can simultaneously constrain demand and reveal the value that beneficiaries place on the service. If those benefiting from a service do not bear the full social cost of the service, they may seek to have the government provide more of the service than is economically efficient. User fees may also foster production efficiency by increasing awareness of the costs of publicly-provided services, therefore increasing incentives to reducecosts where possible.
Equity	Equity means that everyone pays their fair share, but the definition of fair share can have multiple facets, in part because beneficiaries and users may not be the same. Under the beneficiary-pays principle, the beneficiaries of a service pay for the cost of providing the service from which they benefit. Under the ability-to-pay principle, those who are more capable of bearing the burden of fees should pay more for the service than those with less ability to pay.
Revenue adequacy	Revenue adequacy is the extent to which the fee collections cover the intended share of costs. It encompasses the extent to which collections may change over time relative to the cost of the program. For the purposes of our work, revenue adequacy also incorporates the concept of revenue stability, which generally refers to the degree to which short term fluctuations in economic activity and other factors affect the level of fee collections.
Administrative burden	This is the cost of administering the fee, including the cost of collection and enforcement, as well as the compliance burden (the administrative costs imposed on the payers of the fee).

Source: GAO-08-386SP.

APPENDIX IV: THE FAIR INFORMATION PRACTICES

The following Fair Information Practices, with some variation, are used by organizations to address privacy considerations in their business practices and are also the basis of privacy laws and related policies in the United States, the European Union, and countries including Australia and New Zealand.

These practices are not precise legal requirements. Rather, they provide a framework of principles for balancing the need for privacy with other public policy interests, such as national security, law enforcement, and administrative efficiency.[1]

Table 6. The Fair Information Practices

Principle	Description
Collection limitation	The collection of personal information should be limited, should be obtained by lawful and fair means, and, where appropriate, with the knowledge or consent of the individual.
Data quality	Personal information should be relevant to the purpose for which it is collected, and should be accurate, complete, and current as needed for that purpose.
Purpose specification	The purposes for the collection of personal information should be disclosed before collection and upon any change to that purpose, and its use should be limited to those purposes and compatible purposes.
Use limitation	Personal information should not be disclosed or otherwise used for other than a specified purpose without consent of the individual or legal authority.
Security safeguards	Personal information should be protected with reasonable security safeguards against risks such as lossor unauthorized access, destruction, use, modification, or disclosure.
Openness	The public should be informed about privacy policies and practices, and individuals should have ready means of learning about the use of personal information.
Individual participation	Individuals should have the following rights: to know about the collection of personal information, to access that information, to request correction, and to challenge the denial of those rights.
Accountability	Individuals controlling the collection or use of personal information should be accountable for taking steps to ensure the implementation of these principles.

Source: Organization for Economic Cooperation and Development.

Appendix V: Mileage-based User Fee Simulation Model

An Overview

This appendix describes the algorithm of the simulation model that we used to estimate mileage-based user fees (mileage fees) for various hypothetical scenarios with different target revenues.

Methodology

The simulation model comprises four parameters and four variables (see table 7), and the algorithm of the simulation mainly involves repetition of three steps of computation until a desired target revenue is achieved. While the values of the parameters do not change, the variables in the model change from step to step. As described in appendix I, we calibrated the four parameters using existing literature, government data or other credible data sources. We then computed values of variables as formulated in the following equations. For expositional purpose, the algorithm and its steps are shown as a flow chart in figure 14.

Table 7. Description of Key Parameters and Variables Used in Simulation Analysis

Description	Symbol	Remark
Parameters		
Vehicle miles traveled (VMT) elasticity to fuel price	β	Measures the extent VMT would change in response to change in fuel price
Fuel tax (dollars per gallon)	*tf*	Federal tax on gasoline or diesel fuel, and does not include state or other fuel taxes
Average fuel efficiency of vehicles (miles per gallon)	*FE*	
Per-mile fuel tax	μ	Measures how much the current federal fuel tax costs per mile for avehicle with average fuel economy, and is, mathematically, a dividend of tax per gallon (i.e., fuel tax (*tf*)) and average miles per gallon (i.e. fuel economy) and can be expressed as $\mu = tf/FE$
Variables		
Mileage fee (dollars per mile)	Φ	
Fuel revenue (dollars)	Ω	Pre-determined revenue denoted as (Ω^*)
VMT (miles)	*VMT*	
Retail fuel price (dollars per gallon)	*P*	Includes federal and state fuel tax, and is the effective cost of fuel users pay

Source: GAO.

Step One: We first computed the mileage fee rates for passenger vehicles and commercial trucks using the pre-determined target revenue (Ω^*) and

VMT. For example, for the first iteration of the first scenario to replace the fuel tax revenue in 2010, we computed the initial mileage fee rates ($\Phi 0$) using gasoline and diesel tax receipts ($\Omega^* = \Omega 2010$) and total number of vehicle miles traveled in 2010 (VMT^{2010}) as described in equation (1a).

Equation (1a) $$\Phi 0 = \Omega^{2010}/VMT^{2010};$$

Next, we computed the new retail fuel price ($P1$). Since fuel price is expressed on a per gallon basis, we converted the mileage fee ($\Phi 0$) into a new per-gallon fuel tax (mathematically, this can be expressed as ($\Phi 0*FE$)). We first subtracted the existing per-gallon fuel tax (tf) from the 2010 fuel price and added back the new per-gallon fuel tax ($\Phi 0*FE$) to get to the new fuel retail price.

For example, using the retail fuel price for 2010, the new retail fuel price was computed as follows:

Equation (1b) $$P1=[P^{2010} - (tf - \Phi 0*FE)];$$

Step Two: We used the new retail fuel price ($P1$) computed in step one to compute the number of total miles traveled ($VMT1$) as we assumed that users would respond to the fuel price change by changing the amount of driving. For example, using the VMT^{2010} for fiscal year 2010, the new $VMT1$ was derived as follows:

Equation (2) $$VMT1=VMT^{2010} *(1+\beta*(P1-P^{2010})/P^{2010})$$

Step Three: Lastly, we computed the fuel tax revenue ($\Omega 1$) by multiplying the new $VMT1$ and mileage fee ($\Phi 0$) as follows.

Equation (3) $$\Omega 1 = VMT1* \Phi 0;$$

The algorithm repeats itself by feeding the new tax revenue ($\Omega 1$) into equation (1a) until the final mileage fee, when multiplied by the final VMT, generates a pre-determined target revenue (Ω^*) within a set level of approximation.

Equation (1a) $$\Phi 1 = \Omega 1/VMT^{2010}$$

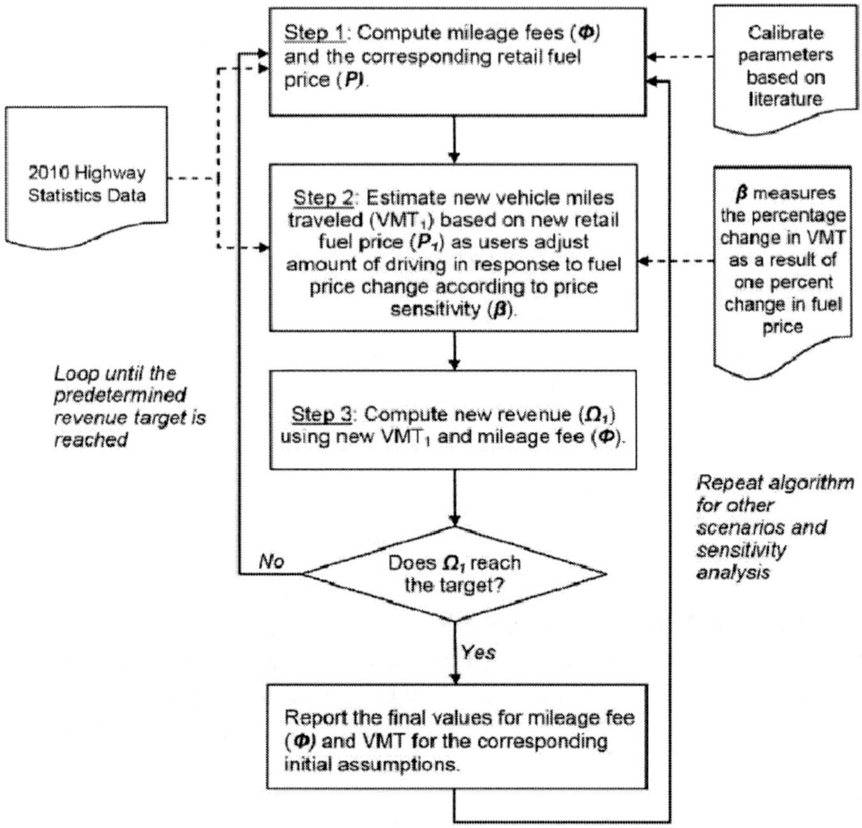

Source: GAO.

Figure 14. Overview of the Algorithm of Computing Mileage Fees to Generate Predetermined Revenue.

Assumptions and Limitations

Because the simulation model focused on what mileage fee rates were required to generate different target revenues, we did not account for the following factors and conditions, which would significantly affect the mileage fees:

1. There were only two flat mileage fees, one for passenger vehicles and one for commercial trucks. We did not account for any externalities

such as air pollution or congestion costs, nor did we model the mileage fees to vary with different types of vehicles or income levels to achieve social and environmental goals.

2. We assumed that vehicles fall in two broad categories marked by two average fuel-efficiency levels. In reality, there is substantial variation in fuel efficiency of vehicles within each category.
3. We assumed a uniform elasticity (i.e., response to a price change within each category of vehicles) even though owners of different types of vehicles have varying degrees of price sensitivities.
4. We assumed users only change the extent of their driving (VMT) in response to changes in fuel price as a result of a change in tax. In reality, users' responses could also include behavior changes such as switching to public transit or carpooling and this would also have an effect on the mileage fees.
5. We did not account for factors other than fuel taxes that might affect VMT. These factors such as the strength of economy or higher required fuel economy for vehicles could significantly influence people's driving. To the extent that these non- tax factors affect VMT, mileage fees for passenger vehicles and commercial trucks would change to meet the target revenues in the hypothetical scenarios.

APPENDIX VI: RESULTS OF GAO SURVEY OF STATE DEPARTMENTS OF TRANSPORTATION ABOUT MILEAGE FEES

To identify states' views on addressing future revenue demands using mileage fees, also known as vehicle miles traveled (VMT) user fees, we surveyed the departments of transportation in all 50 states and the District of Columbia from March 22 through April 24, 2012 and received a 100 percent response rate.

VMT User Fee Evaluation Efforts and Potential Challenges

1. Has your state DOT taken or planned to take any of the following steps to evaluate VMT user fees? (Check one per row.)

Question	Response				
	Yes	No, but plan to do so in the next 12 months	No, and have no plans in the next 12 months	Don't Know	Total responses
a. Reviewed existing research	36	3	11	1	51
b. Conducted a review of the technologies and systems available to administer a VMT user fee program	16	6	27	2	51
c. Conducted economic analysis on the viability of a VMT system	5	5	39	2	51
d. Conducted research (e.g., survey or focus groups) to gauge the public's potential acceptance	3	7	41		51
e. Conducted a pilot project to test a VMT user fee system	3	4	44		51
f. Conducted specific research to evaluate VMT user fees for commercial trucks	7	3	40	1	51
g. Participated in research with other states to evaluate VMT user fees	14	5	30	2	51
h. Shared information with other states related to VMT user fees	17	5	28	1	51
i. Other	6	1	6	8	21

2. Did your state legislature, governor, or transportation commission direct your state DOT to evaluate VMT user fees? (Check one per row.)

Question			Response	
	Yes	No	Not applicable/ Don't know	Total responses
a. State legislature directed state DOT	8	39	3	50
b. Governor's office directed state DOT	3	44	3	50
c. State transportation commission directed state DOT	3	43	4	50
d. Other	7	12	6	25

3. How much of a challenge might the following issues present to developing a VMT user fee program in your state? (Check one per row.)

Question	Response						
	Very great challenge	Great challenge	Moderate challenge	Some challenge	No challenge	Don't know	Totalresponses
a. Addressing technological issues	8	13	21	4	2	3	51
b. Addressing privacy concerns	23	22	1	2	1	2	51
c. Educating the public about the viability of the current gas tax to meet funding demands	18	9	15	7	2		51
d. Obtaining public support for a VMT user fee program	36	11	2			2	51
e. Obtaining support from elected officials for a VMT user fee program	30	12	6	1		2	51
f. Addressing equity concerns for how different groups (e.g., rural, urban, low-income drivers) could be affected	11	19	14	3		4	51
g. Administrative costs of implementing VMT program (e.g., collection of fees, enforcement, compliance)	14	15	13	5		4	51
h. Developing the technical capacity or expertise to implement a VMT program	7	9	26	7		2	51
i. Legal barriers to implementing a VMT program in state (e.g., required changes to state's Constitution)	9	15	9	5	2	11	51
j. Obtaining funding necessary to evaluate or test a VMT system	9	10	17	7	2	6	51
k. Other	3		1			9	13

Potential Federal Actions to Develop VMT User Fee Programs

4. To what extent would your state DOT support or oppose the federal government taking the following actions toward the development of future VMT user fee systems? (Check one per row.)

Question	Response						
	Strongly support	Support	Neither support nor oppose	Oppose	Strongly oppose	Don't know	Totalresponses
a. Establishing performance & interoperability standards for the development of VMT user fee systems	12	21	9	2	2	5	51
b. Establishing a process to certify VMT systems developed by the private sector	12	19	13	1	1	5	51
c. Establishing a framework for states to coordinate research and testing of VMT user fee systems	15	23	8	1	1	3	51
d. Providing Information to states on appropriate VMT user fee rates	9	21	13	3	2	3	51
e. Providing incentives to states to pilot or implement VMT user fee systems	10	21	15	2	1	2	51
f. Conducting research on the cost and benefits of VMT systems	13	27	7		1	3	51
g. Developing a communication plan to educate the public about transportation funding issues	19	21	6		1	3	50
h. Other	3	1	1		1	9	15

5. Would your state DOT support or oppose federally-led field tests to evaluate VMT user fees for the following types of vehicles? (Check one per row.)

Question	Response						
	Strongly support	Support	Neither support nor oppose	Oppose	Strongly oppose	Don't know	Totalresponses
a. Passenger vehicles	5	24	16		1	5	51
b. Commercial trucks	9	21	15		1	5	51
c. Electric vehicles	9	25	13			4	51

6. How important or unimportant is it that the following options are included in future VMT field tests? (Check one per row.)

Question	Response						
	Very important	Important	Neither important nor unimportant	Unimportant	Very unimportant	Don't know	Totalresponses
a. Low tech options to collect mileage data (e.g., odometer readings taken during vehicle registration or inspections)	12	27	4	3	1	4	51
b. Systems which provide drivers with multiple options for submitting mileage data (e.g., on-board GPS units, odometer readings, self-reporting)	10	34	3	1		3	51
c. Systems with variable VMT fees based on vehicle fuel efficiency or weight	10	24	9	1	1	6	51
d. Systems with variable pricing elements based on time-of-day or congestion levels	12	17	11	4	2	5	51
e. Systems that provide users the option of paying fuel taxes in lieu of a VMT fee	8	22	12	4	1	4	51

(Continued)

Question	Response						
	Very important	Important	Neither important nor unimportant	Unimportant	Very unimportant	Don't know	Totalresponses
f. Systems available through the marketplace (e.g., factory-installed devices in new vehicles, commercially available GPS units, or smartphone applications)	12	30	5		1	3	51
g. Evaluation of equity of VMT fees across different populations of drivers (e.g., fairness for rural, urban, low-income drivers)	20	20	7	1		2	50
h. Other	4	2	1			8	15

Identifying Additional Surface Transportation Revenue Sources

7. How important or unimportant is it that an alternative funding mechanism be identified by either your state or the federal government to meet surface transportation revenue needs in the next ten years? (Check one per row.)

Question	Response						
	Very important	Important	Neither important nor unimportant	Unimportant	Very unimportant	Don't know	Totalresponses
a. State government	32	17				2	51
b. Federal government	37	13				1	51

8. What level of priority should be placed on the following funding mechanisms to meet future surface transportation revenue needs? (Check one per row.)

Question	Response						
	Highest priority	High priority	Moderate priority	Low priority	Lowest priority	Don't know	Totalresponses
a. Increasing state fuel taxes	7	11	10	8	8	5	49
b. Introducing state-led VMT user fee	1	9	15	12	3	9	49
c. Increasing state vehicle fees (e.g., vehicle registration or emissions testing fee)	2	9	18	6	8	6	49
d. Increasing the use of tolling (e.g., tolls on particular facilities)	8	9	15	8	5	4	49
e. Increasing federal fuel taxes	10	13	11	5	5	5	49
f. Introducing federally-led VMT user fee	10	11	13	5	1	9	49
g. Other	4	1				10	15

9. How likely or unlikely is it that your state will introduce a VMT user fee program for the following types of vehicles in the next ten years? (Check one per row.)

Question	Response						
	Very likely	Likely	As likely asunlikely	Unlikely	Very unlikely	Don't know	Totalresponses
a. Passenger vehicles	1	3	15	14	11	7	51
b. Commercial trucks	1	2	16	15	9	7	50
c. Electric vehicles	3	4	15	11	9	9	51

Additional Comments

10. What additional comments do you have on VMT user fees?
[Open-ended]

Contact Information

11. What is the name, title, telephone number, and e-mail address of the person who primarily completed this survey? We might contact this person if we have follow-up questions.

a. Name
b. Title
c. Department
d. Telephone
e. E-mail

Completion

12. Please check one of the options below. Clicking on "Completed" indicates that your answers are official and final.

	Completed	Not Completed	Total responses
Your answers will not be used unless you have done this. (Check one.)	51		51

Source: GAO.

End Notes

[1] The Highway Trust Fund is an account established by law to hold federal highway user tax receipts (e.g., receipts for federal excise taxes on fuel and other taxes on commercial trucks) that are dedicated for highway and transit related purposes. It is composed of two accounts: the highway account and the mass transit account.

[2] In fiscal year 2008, about $8 billion was transferred from the general fund to the highway account (Pub. L. No. 110-318, § 1(a)(4), 122 Stat. 3532 (Sept. 15, 2008)); in fiscal year 2009 the transfer was $7 billion (Pub. L. No. 111-46, §1, 123 Stat 1970 (Aug. 7, 2009)); and $14.7 billion was transferred to extend highway programs to December 31, 2010, and $4.8 billion was transferred to the mass transit account (Hiring Incentives to Restore Employment Act, Pub. L. No. 111-147, § 442, 124 Stat. 71, 94 (Mar. 18, 2010)). For fiscal year 2013, $6.2 billion is appropriated to the highway account; for fiscal year 2014, $10.4 billion is appropriated to the highway account and $2.2 billion is appropriated to the mass

transit account (Moving Ahead for Progress in the 21st Century Act (MAP-21), Pub. L. No. 112-141, § 4025, 126 Stat. 405, 864 (July 6, 2012)).

[3] GAO, *Highway Trust Fund: All States Received More Funding Than They Contributed in Highway Taxes from 2005 to 2009*, GAO-11-918 (Washington, D.C.: Sept. 8, 2011).

[4] MAP-21.

[5] GAO, *High-Risk Series: An Update*, GAO-11-278 (Washington, D.C.: February 2011).

[6] Congressional Budget Office, *Alternative Approaches to Funding Highways,* Pub. No. 4090 (Washington, D.C., March 2011).

[7] GAO, *Federal User Fees: A Design Guide,* GAO-08-386SP (Washington, D.C.: May 29, 2008).

[8] National Surface Transportation Policy and Revenue Study Commission, *Transportation for Tomorrow* (Washington, D.C., December 2007); National Surface Transportation Infrastructure Financing Commission, *Paying Our Way: A New Framework for Transportation Finance* (Washington, D.C., February 2009).

[9] Pub. L. No. 109-59, § 1919 (a), (d), 119 Stat. 1144, 1479-1480 (2005).

[10] Federal fuel taxes are not directly paid by highway users but are reflected in retail fuel prices. Oil companies typically pay a per-gallon tax on fuels at the point of distribution, and these costs become part of the purchase price paid by highway users.

[11] This includes $24.8 billion and $9.1 billion in federal gasoline and diesel taxes, respectively, and $2.8 billion in truck taxes. This amount does not include the $19.5 billion in general fund revenues deposited into the Highway Trust Fund or the $1.2 billion in rescissions and transfers to other funds in 2010.

[12] Federal truck taxes include: (1) truck and trailer sales tax of 12 percent of retailer's sale price for tractors and trucks over 33,000 pounds gross vehicle weight and trailers over 26,000 pounds gross vehicle weight; (2) a heavy-vehicle use tax for trucks weighing 55,000 pounds or more that ranges from $100 to a maximum of $550 per year based on weight; and (3) a tire tax for the purchase of tires, based on tire size and weight, of 9.45 cents for each 10 pounds of the maximum rated load capacity over 3,500 pounds.

[13] GAO, *Transportation: Key Issues and Management Challenges*, GAO-12-581T (Washington, D.C.: Mar. 29, 2012).

[14] In addition, drivers also pay an average of 23.3 cents per gallon in state gasoline taxes; rates range from 8 cents per gallon in Alaska and Georgia to 39.2 cents per gallon in North Carolina. Some states also levy additional taxes and fees on the sale of gasoline and diesel fuel, which increases the effective tax rate per gallon.

[15] 49 C.F.R. Parts 531, 533, as amended by *Light-Duty Vehicle Greenhouse Gas Emission Standards and Corporate Average Fuel Economy Standards.* 75 Fed. Reg. 25324 (2010).

[16] *2017 and Later Model Year Light-Duty Vehicle Greenhouse Gas Emissions and Corporate Average Fuel Economy Standards,* 77 Fed. Reg. 62624, 62627 (Oct. 15, 2012), errata issued, 77 Fed. Reg. 64051 (Oct. 18, 2012), 77 Fed. Reg. 68070 (Nov. 15, 2012).

[17] GAO-11-918.

[18] The National Household Transportation Survey (NHTS) includes information on the household-based vehicle fleet, including the fuel efficiency of each vehicle, annual miles of vehicle use, the resulting amount of gallons of gasoline consumed, and the average cost of gasoline at the household location during the interview period. The most recent NHTS was collected from April 2008 through April 2009. The average cost for a gallon of gasoline during that 13-month period was $2.96, although prices reached as high as $4.00 in that period. FHWA, *2009 National Household Travel Survey* (Washington, D.C., June 2011).

[19] Single-unit trucks have a single frame, two axles, and at least six tires or a gross vehicle weight rating exceeding 10,000 pounds.

[20] From 1958 to 1960, AASHTO conducted road tests to determine the relationships between axle weights and pavement wear. The tests showed, for example, that an axle weight of 30,000 pounds causes 8 times more pavement damage than an axle weight of 18,000 pounds. The relationships developed from these tests are still used today to attribute pavement wear to various vehicle types.

[21] In 1994, GAO recommended that FHWA conduct an updated highway cost allocation study utilizing data on the relationship between axle loads and pavement damage to determine whether all highway users were paying their fair share of federal highway costs. GAO, *Highway User Fees: Updated Data Needed to Determine Whether All Users Pay Their Fair Share*, GAO/RCED-94-181 (Washington, D.C.: June 7, 1994).

[22] FHWA's 2000 highway cost allocation study estimated costs attributable to each vehicle class using a process that considers how physical and operational characteristics of each vehicle class affect the design of various components of the highway system or the rate at which pavements, bridges, and other elements of the highway infrastructure wear out and must be repaired or replaced. According to FHWA, a vast body of research has demonstrated the relationship between axle loads and pavement wear and that heavy axle loadings contribute significantly to costs for rehabilitating and reconstructing pavements.

[23] SAFETEA-LU §§ 1919 (a), (d), 1934, 119 Stat.,1479-1480, 1490.

[24] The final report was prepared for the U.S. DOT in 2011 and delivered to Congress in April 2012.

[25] FHWA's Value Pricing Pilot Program, as authorized under SAFETEA-LU, § 1604(a), 119 Stat., 1449-1450, encourages the implementation and evaluation of value pricing pilot projects to manage congestion on highways through tolling and other pricing mechanisms.

[26] Although these countries refer to their programs by different names and may charge their user fees based on kilometers driven, rather than miles driven, we refer to these programs generally as mileage fee programs in this report.

[27] GAO-08-386SP.

[28] Recent research by the National Cooperative Highway Research Program (NCHRP) suggests that the administrative costs of the collection of state fuel taxes are about 1 percent of the revenues collected. See NCHRP, *Costs of Alternative Revenue-Generation Systems, Report 689* (Washington, D.C., 2011).

[29] CBO, *Alternative Approaches to Funding Highways,* Pub. No. 4090 (Washington, D.C., March 2011).

[30] See app. IV for information on the Fair Information Practices.

[31] GAO, *Privacy: Alternatives Exist for Enhancing Protection of Personally Identifiable Information*, GAO-08-536 (Washington, D.C.: May 19, 2008).

[32] The five U.S. pilot projects we reviewed were the Iowa study; the Oregon Road User Fee Pilot Program; the Puget Sound Regional Council's Traffic Choices Study; the Minnesota DOT Road Use Test; and the Nevada DOT VMT Fee Study. See app. II for a brief summary of each.

[33] The pilot projects using thick-client systems were the Iowa study, Minnesota DOT's Road Use Test, and Oregon's Road User Fee Pilot Program, which also used pay-at-the pump technology in conjunction with a GPS receiver. The Puget Sound Regional Council used a thin-client GPS system. The Nevada DOT system does not use GPS.

[34] To calculate fees in the vehicle, thick-client units are preloaded with map files that may have fee rates for each jurisdiction (state, county, city) to allow for different charges based on location. The unit determines the total number of miles driven in each jurisdiction to calculate the total fee.

[35] GPS receivers receive signals but do not transmit wide area signals whereas GPS navigational units do both, thus enabling third-party tracking of vehicle movements. The GPS receivers that Oregon DOT used in its pilot program did not have the ability to track vehicle movements.

[36] Although the Oregon DOT concluded that this pilot was successful and could be implemented at a reasonable cost, the department has since revised its technical approach and is working to develop a new system that will not mandate the use of a GPS receiver and will focus on the collection of user fees from electric-powered vehicles, which currently do not pay state or federal fuel taxes.

[37] Electric vehicles are currently exempt from RUC, but New Zealand officials reported that this would change if their numbers became significant.

[38] GAO, *Mobile Device Location Data: Additional Federal Actions Could Help Protect Consumer Privacy*, GAO-12-903 (Washington, D.C.: Sept. 11, 2012).

[39] The options include an on-board unit that does not use GPS and only collects and reports undifferentiated miles; a unit with a GPS receiver that reports miles by location; or a flat annual or biannual tax that does not require the collection of any mileage data and allows the driver to accumulate unlimited miles.

[40] H.R. 5972, § 419, 112th Cong. (2012).

[41] Odometer fraud is an illegal practice that consists of decreasing the number of miles reported by a vehicle's odometer by disabling or turning back the odometer 's reading to appear that a vehicle has lower mileage than what was driven.

[42] The cost per unit for the U.S. vehicle fleet would likely be substantially lower because unit costs would be expected to decrease with more units purchased.

[43] See National Cooperative Highway Research Program, *Costs of Alternative Revenue-Generation Systems, Report 689* (Washington, D.C., 2011).

[44] Switzerland, the Czech Republic, Slovakia, and Austria each have also implemented a commercial truck distance-based user fee system.

[45] Euro to dollar conversions are based an exchange rate of .8198 on July 30, 2012.

[46] Additional charges would apply for trailers, which must also be licensed separately from powered trucks towing them and carry their own distance recorders.

[47] New Zealand to U.S. dollar conversions are based on an exchange rate of 1.32 on July 30, 2012. The total RUC revenues in New Zealand dollars from fiscal years 2007 through 2011 were $4.47 billion in New Zealand dollars. New Zealand transportation officials estimate that in fiscal year 2011, about 10 percent to 15 percent of all commercial truck RUC revenues were generated through the E-RUC devices.

[48] Federal Highway Administration, *Comprehensive Truck Size and Weight Study,* FHWAPL-00-029 (Washington, D.C., August 2000).

[49] The enforcement cameras capture images of each passing truck and its license plate so that the toll operator can verify whether vehicles without on-board units have booked their travel for the appropriate segment of the autobahn.

[50] The mobile enforcement fleet is equipped with short-range communications devices that read a moving truck's on-board unit to test for compliance. To determine whether users without on-board units have registered for their travel, German mobile transport officers can query the toll operator's database of registered users for license plate data to check compliance.

[51] Data on the start-up costs are proprietary to the toll operator and were not available for this report. However, those costs, and the costs of operating the system are recouped through annual, performance-based contractual fees paid by the German government to the toll operator.

[52] According to German transportation officials, these programs are intended to offset the costs of compliance to German trucking firms to upgrade to lower emission vehicles, to train drivers to comply with the system, and a small program to enhance vehicle safety and the environment.

[53] These "hubodometers" are removable mechanical mileage tracking devices that can be mounted on trailers, which are not equipped with odometers, so these vehicles can be licensed and charged.

[54] Unlike the arrangement that Germany has with its toll operator, the New Zealand government pays nothing to the private sector companies that implement the E-RUC system as its revenues are generated from the sale of subscription services to trucking firms.

[55] New Zealand government officials reported that the 4 percent evasion rate reflects the former RUC rules, which were changed in August 2012 in order to reduce noncompliance. The RUC system formerly charged on the basis of actual weight as nominated by vehicle operators, but this was changed in order to improve compliance and simplify administration. As of August 2012, the RUC system charges all vehicles based on their maximum gross-laden weight regardless of whether the vehicle is fully loaded or empty.

[56] See app. IV for information on the Fair Information Practices. The German Motorway Toll Act for Heavy Goods Vehicles (2002) §§ 4(2) and 7(2) establishes limits on the specific data that may be collected, processed, and used for the collection and enforcement of tolls in the HGV system. The New Zealand Road Use Act of 2012, Schedule 1, § 45(3) establishes the duties of E- RUC system providers relating to the collection, use and disclosure, and management of RUC data.

[57] Four states (Oregon, New York, Kentucky, and New Mexico) currently levy weight-distance taxes on commercial trucks. Two of these states, Oregon and New York, have considered transitioning to an electronic collection of these user fees. In Oregon, all trucks over 26,000 pounds must register their highest operating weight for each truck combination and self-report the miles traveled in Oregon by each combination. The weight-mile tax rates vary in relation to axle load and weight to incentivize the use of truck combinations that do less damage to roadways. The taxes are paid on a monthly or quarterly basis. New York's ton-mile tax charges trucks different rates based on whether they are hauling cargo (laden weight) or are empty (unladen weight).

[58] Delcan Corporation, Calmar Telematics, and the Greater Buffalo Niagara Regional Transportation Council, *A Practical Approach to Truck VMT Fees, Final Report* (April 2011).

[59] 49 U.S.C. § 31705, referring to and permitting states to require fuel use tax reporting requirements conforming to IFTA.

[60] 49 U.S.C. § 31704, referring to and restricting states not participating in IRP.

[61] IFTA is an interstate agreement for the collection and distribution of fuel taxes. IRP is an interstate agreement for the apportionment of registration fees. The Intermodal Surface Transportation and Efficiency Act of 1991 encouraged states to participate in these agreements. Under these agreements, states collect data on the miles that truck operators travel in their state and on registered truck weights.

[62] The other federal taxes paid by commercial trucks include the federal heavy vehicle use tax and the federal tire, truck, and trailer excise taxes, which generated $2.8 billion in fiscal year 2010.

[63] Federal fuel tax receipts totaled almost $34 billion in fiscal year 2010. We attributed the $24.8 billion in federal gasoline tax receipts deposited in the Highway Trust Fund to passenger vehicles and the $9.1 billion generated through the federal diesel fuel tax to commercial

trucks. Our model does not consider the small portion of the U.S. passenger vehicle fleet powered by diesel fuel. According to the Environmental Protection Agency, less than 1 percent of all model year 2010 light duty vehicles in the United States ran on diesel fuel and, as result, the amount of federal diesel fuel tax receipts contributed to the Highway Trust Fund by passenger vehicles is minimal. It also does not consider that small portions of federal gasoline fuel tax receipts are estimated to be derived from use in motorboats and small engines, such as lawnmowers and chain saws, and transferred from the Highway Trust Fund to other nonhighway accounts.

[64] The $53.5 billion revenue target in this scenario represents the total budget authority for FHWA and the Federal Transit Administration. This amount does not include the $27.5 billion in general fund revenues identified through the American Recovery and Reinvestment Act of 2009 for federal highway infrastructure projects (Pub. L. No. 111-5, Title XII, 123 Stat. 115, 206 (Feb. 17, 2009).

[65] MAP-21 authorized funding for fiscal years 2013 and 2014. According to FHWA and the Federal Transit Administration, the total amount of funds authorized for highway and transit programs in these years is more than $103 billion.

[66] In February 2009, the commission reported that annual federal highway and transit spending of $59 billion and $19 billion, respectively, would be necessary to maintain the existing conditions and performance of our nation's surface transportation system. National Surface Transportation Infrastructure Financing Commission, *Paying Our Way: A New Framework for Transportation Finance* (Washington, D.C., February 2009). After adjusting this figure to reflect 2010 dollars and subtracting the $2.8 billion that commercial trucks contributed to the Highway Trust Fund in other taxes in fiscal year 2010, this target is equal to $78.4 billion in federal fuel tax receipts.

[67] These estimates were based on existing academic literature that has attempted to identify the range of costs of implementing a mileage fee system. For details on how we modeled mileage fee rates that included administrative costs, see app. I.

[68] The number of miles traveled varies widely among different types of commercial trucks. Our projections are based on the statistics compiled using two categories of commercial trucks in FHWA Highway Statistics as well as fuel efficiency estimates for different types of single-unit and combination trucks. See National Academy of Sciences, *Technologies and Approaches to Reducing the Fuel Consumption of Medium- and Heavy-Duty Vehicles* (Washington, D.C., 2010).

[69] These projections simply apply a flat commercial truck rate to two truck types. In theory, heavier trucks would need to pay higher mileage fees than lighter trucks because they are responsible for significantly more road damage than lighter trucks.

[70] GAO-08-386SP.

[71] GAO, *Surface Freight Transportation: A Comparison of the Costs of Road, Rail, and Waterways Freight Shipments That Are Not Passed on to Consumers,* GAO-11-134 (Washington, D.C.: Jan. 26, 2011).

[72] According to FHWA, passenger vehicles represent the other 96 percent of the U.S. vehicle fleet and are responsible for 59.7 percent of road costs incurred. FHWA's 2000 addendum to its *1997 Highway Cost Allocation Study* noted that this determination is based on federal spending patterns, and FHWA officials said that estimates that illustrate different vehicle types' responsibility for highway costs irrespective of federal spending are not available.

[73] In fiscal year 2010, 67.6 percent of Highway Trust Fund user fee revenues were generated through the federal gasoline tax.

[74] GAO/RCED-94-181.

[75] Conversely, individual passenger vehicles cause only very limited amounts of wear on the road, and the variation in wear among different light vehicles of different weights is also minimal.

[76] Failing to charge for the additional wear caused by the heaviest trucks on the highways can result in a distortion of the competitive environment by making it appear that these trucks are a less expensive means to transport goods.

[77] GAO/RCED-94-181.

[78] Converting the adjusted mileage fee rates to per-gallon charges would translate to significantly higher federal diesel fuel tax rates. For example, an adjusted federal diesel fuel tax of 45 cents per gallon would be needed under the scenario to meet current spending levels and a diesel fuel tax of 69.3 cents per gallon would be necessary under the scenario to maintain existing conditions and performance levels.

[79] GAO/RCED-94-181.

[80] GAO-08-386SP.

[81] New Jersey was the only state DOT to indicate that it does not know whether it is important that an alternative federal funding mechanism be identified in the next 10 years in order to meet surface transportation revenue needs.

[82] Department of Energy, *One Million Electric Vehicles by 2015: February 2011 Status Report* (Washington, D.C., 2011).

[83] Texas Transportation Institute, *Exploratory Study: Vehicle Mileage Fees in Texas* (College Station, Tex., January 2011).

[84] FHWA, *Potential Research and Field Trials Related to Mileage-Based User Fees*, Report to Congress (Washington, D.C., March 2011).

[85] GAO, *Energy Policy: Options to Reduce Environmental and Other Costs of Gasoline Consumption*, GAO/RCED-92-260 (Washington, D.C.: Sept. 17, 1992).

End Notes for Appendix I

[1] Pub. L. No. 109-59, § 1919 (a), (d), 119 Stat. 1144, 1479-1480 (2005).

[2] See appendix II for a summary of U.S.-based pilot projects as of June 2012.

[3] FHWA, *Highway Statistics 2010,* Table FE -10 (Washington, D.C., 2012).

[4] In 2010, FHWA reported that the amounts of average annual spending needed to maintain the conditions and performance of the nation's surface transportation systems to be $101 billion for highways and $18 billion for transit. FHWA, *2010 Status of the Nation's Highways, Bridges, and Transit: Conditions & Performance Report to Congress* (Washington, D.C., 2012). The difference between these figures and our scenario 3 revenue target is explained, in part, by: (1) the $78.4 billion target consists entirely of federal revenues while the $119 billion in the Condition and Performance report represents all levels of government and (2) both figures are average annual revenues that use different projected timeframes.

[5] The National Surface Transportation Infrastructure Financing Commission's 2009 report stated that annual federal highway and transit spending requirements to maintain existing conditions and performance were $59 billion and $19 billion (in 2008 dollars), respectively, were necessary. Adjusted to reflect 2010 dollars and not including the other excise taxes and fees levied on commercial trucks, this is equal to $78.4 billion, per FHWA.

[6] FHWA, *Addendum to the 1997 Highway Cost Allocation Study Final Report*, Table 4, (Washington, D.C., May 2000).

[7] In computing the cost responsibility ratio for all passenger vehicles, FHWA's 2000 update to its *1997 Highway Cost Allocation Study* included buses. However, in FHWA's *Highway Statistics 2010*, the all light-duty vehicles category did not include buses. We decided that the difference is small as VMT by buses on urban and rural roads (13,789 million) was only 0.5 percent of that of all light-duty vehicles. See FHWA, *Addendum to 1997 Highway Cost Allocation Study,* Table 4; and FHWA, *Highway Statistics 2010*, Table VM-1.

[8] Our model does not consider that small portions of federal gasoline fuel tax receipts are estimated to be derived from use in motorboats and small engines, such as lawnmowers and chain saws, and transferred from the Highway Trust Fund to other non-highway accounts.

[9] U.S. Environmental Protection Agency, *Light-Duty Automotive Technology, Carbon Dioxide Emissions, and Fuel Economy Trends, 1975-2010,* page viii, EPA-420R-12-001, (Washington, D.C., March 2012).

[10] A direct comparison with the existing literature is not appropriate as these studies used diverse methodologies, assumptions or data. For example, our model did not account for external costs such as vehicle emissions and mileage fee rates that account for these costs would normally be higher than fees designed to only replace current revenues. As such, we expected our results to be at lower end of the existing estimates. For example, Parry and Small (2005) estimated an optimal mileage fee to account for all measurable external costs to be 14 cents per mile, which is significantly higher than the rates in our simulation. See Parry, Ian and Kenneth Small, Does Britain or the United States Have the Right Gasoline Tax?, *American Economic Review*, Vol. 95. No.4, page 1283 (2005); Congressional Budget Office, *Alternative Approaches to Funding Highways,* Pub. No. 4090 (Washington, D.C., March 2011); FHWA, *Addendum to the 1997 Highway Cost Allocation Study*, Table 13, (Washington, D.C., May 2000); Delcan Corporation, Calmar Telematics, and the Greater Buffalo Niagara Regional Transportation Council, *A Practical Approach to Truck VMT Fees, Final Report* (April 2011); Parry, Ian, How Should Heavy-Duty Trucks Be Taxed?, *Resources for the Future Discussion Paper*, RFF DP 06-23 (Washington, D.C., 2006).

[11] National Academy of Sciences, *Technologies and Approaches to Reducing the Fuel Consumption of Medium- and Heavy-Duty Vehicles* (Washington, D.C., 2010).

End Notes for Appendix III

[1] GAO, *Federal User Fees: A Design Guide,* GAO-08-386SP (Washington, D.C.: May 29, 2008).

End Notes for Appendix IV

[1] GAO, *Privacy: Alternatives Exist for Enhancing Protection of Personally Identifiable Information*, GAO-08-536 (Washington, D.C.: May 19, 2008).

In: Mileage Fees and the Highway Trust Fund ISBN: 978-1-62618-829-7
Editor: Yoan Bergeron

Chapter 2

HIGHWAY TRUST FUND: ALL STATES RECEIVED MORE FUNDING THAN THEY CONTRIBUTED IN HIGHWAY TAXES FROM 2005 TO 2009

United States Government Accountability Office

WHY GAO DID THIS STUDY

Federal funding for highways is provided to the states mostly through a series of grant programs known as the Federal-Aid Highway Program, administered by the Department of Transportation's (DOT) Federal Highway Administration (FHWA). In 2005, the Safe, Accountable, Flexible, Efficient Transportation Equity Act: A Legacy for Users (SAFETEA-LU) authorized $197.5 billion for the Federal-Aid Highway Program for fiscal years 2005 through 2009. The program operates on a "user pay" system, wherein users contribute to the Highway Trust Fund through fuel taxes and other fees. The distribution of funding among the states has been a contentious issue. States that receive less than highway users contribute are known as "donor" states and states that receive more than users contribute are known as "donee" states.

GAO was asked to examine for the SAFETEA-LU period (1) how contributions to the Highway Trust Fund compared with the funding states received, (2) what provisions were used to address rate-of-return issues across states, and (3) what additional factors affect the relationship between contributions to the Highway Trust Fund and the funding states receive. To conduct this review, GAO obtained and analyzed data from FHWA, reviewed FHWA and other reports, and interviewed FHWA and DOT officials.

GAO is not making any recommendations. DOT reviewed a draft of this report and provided technical comments, which we incorporated as appropriate.

WHAT GAO FOUND

From 2005 to 2009, every state received more funding for highway programs than they contributed to the Highway Account of the Highway Trust Fund. This was possible because more funding was authorized and apportioned than was collected from the states, and the fund was augmented with about $30 billion in general revenues since fiscal year 2008. If the percentage of funds states contributed to the total is compared with the percentage of funds states received (i.e., relative share), then 28 states received a relatively lower share and 22 states received a relatively higher share than they contributed. Thus, depending on the method of calculation, the same state can appear to be either a donor or donee state.

The Equity Bonus Program was used to address rate-of-return issues. It guaranteed a minimum return to states, providing them with about $44 billion. Nearly all states received Equity Bonus funding, and about half received a significant increase—at least 25 percent—over their core funding.

The infusion of general revenues into the Highway Trust Fund affects the relationship between funding and contributions, as a significant amount of highway funding is no longer provided by highway users. Additionally, using rate of return as a major factor in determining highway funding poses challenges related to performance and accountability in the highway program; in effect, rate-of-return calculations override other considerations to yield a largely predetermined outcome—that of returning revenues to their state of origin. Because of these and other challenges, funding surface transportation programs remains on GAO's High-Risk list.

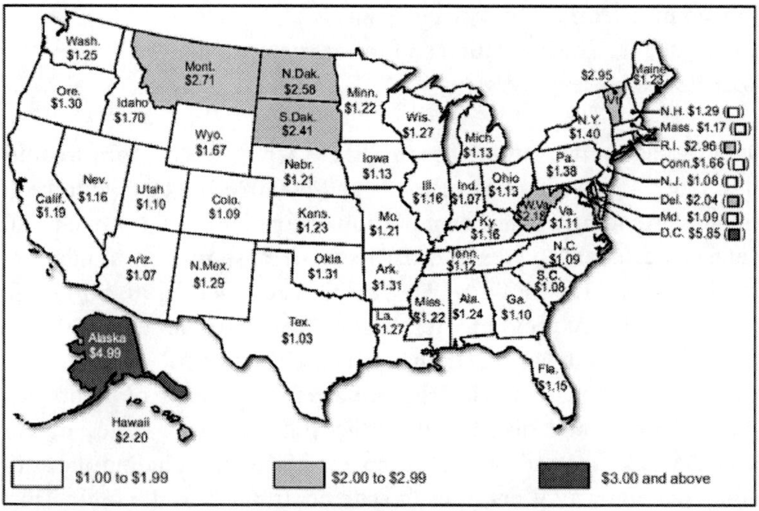

Sources: GAO analysis of FHWA data; Map Resources (map).

States' Return per Dollar Contributed to the Highway Account of the Highway Trust Fund, Fiscal Years 2005-2009.

ABBREVIATIONS

CRS	Congressional Research Service
DOT	Department of Transportation
FHWA	Federal Highway Administration
SAFETEA-LU	Safe, Accountable, Flexible, Efficient Transportation Equity Act: A Legacy for Users
TEA-21	Transportation Equity Act for the 21st Century

September 8, 2011

The Honorable Nick J. Rahall II
Ranking Member
Committee on Transportation and Infrastructure
House of Representatives

The Honorable Peter A. DeFazio
Ranking Member

Subcommittee on Highways and Transit
Committee on Transportation and Infrastructure
House of Representatives

Over decades, the nation has built a vast highway infrastructure that includes about 4 million miles of roads and 600,000 bridges, with the federal government providing a significant portion of funding for this system. In 2005, the Safe, Accountable, Flexible, Efficient Transportation Equity Act: A Legacy for Users (SAFETEA-LU)[1] authorized over $190 billion for the Federal-Aid Highway programs for fiscal years 2005 through 2009, and the Highway Trust Fund has been the principal source of funding for these programs. These funds are primarily collected from taxes on motor fuel and truck-related items and distributed to the states using a series of complex formulas that take into account a number of factors, including the estimated share of taxes highway users in each state contributed to the fund. This "user pay" system, wherein users contribute to the building and upkeep of the system, has generated debate about the funding distribution among states, and states have taken a strong interest in their respective rate of return on contributions.

States that receive less than the estimated contributions paid in-state by highway users are known as "donor" states, while states that receive more than their estimated contributions are known as "donee" states. In June 2010, we reported that nearly all states received more highway funding than they contributed for fiscal years 2005 through 2008.[2] This occurred because, overall, more funding was authorized and apportioned than was collected in-state from highway users (states' contributions). The account was supplemented by about $29.7 billion from general revenue funds into the Highway Trust Fund since fiscal year 2008. The infusion of general revenue funds has changed the "user pay" principle and complicated the link between highway taxes and highway funding. The issue of states' rates of return and donor-donee states continues to be the subject of congressional debate, including during reauthorization of SAFETEA-LU.[3]

To better understand the relationship between states' contributions to the Highway Trust Fund and the amount of federal funding states received, at your request, we updated our June 2010 report and examined:

- the amount of revenue contributed to the Highway Trust Fund Highway Account compared with the funding states received during the SAFETEA-LU period;

- the provisions in place during the SAFETEA-LU period to address rate-of-return issues across states, and how they affected the highway funding states received; and
- additional factors that affected the relationship between contributions to the Highway Trust Fund and the funding states receive.

This update of our June 2010 report includes fiscal year 2009 data on the states' contributions to the Highway Trust Fund that was made available by Federal Highway Administration (FHWA).[4]

To determine how the amount of revenue states contributed compared to funding states received (states' rates of return), we obtained and analyzed FHWA data, including estimates of payments made into the Highway Account of the Highway Trust Fund, by state, and the actual total apportionments states received during the entire 5-year SAFETEA-LU period. Because different methods of calculating a rate of return can provide different results, we analyzed states' rates of return using four different scenarios.

To determine provisions in place to address rate-of-return issues and their effect on funding, we obtained and analyzed FHWA data on state funding during the SAFETEA-LU period and determined the extent to which rate-of-return provisions increased state funding.

To determine additional factors that affect the relationship between contribution to the Highway Trust Fund and the states' funding, we reviewed FHWA, GAO, Congressional Research Service, and other reports, including our body of work on surface transportation financing and the Highway Trust Fund. For each objective, we also reviewed Department of Transportation (DOT) and FHWA reports and interviewed DOT and FHWA officials. We also obtained information from FHWA on the steps taken to ensure data reliability and determined the data were sufficiently reliable for our purposes.

We conducted this performance audit from June 2011 through September 2011 in accordance with generally accepted government auditing standards. Those standards require that we plan and perform the audit to obtain sufficient, appropriate evidence to provide a reasonable basis for our findings and conclusions based on our audit objectives.

We believe that the evidence obtained provides a reasonable basis for our findings and conclusions based on our audit objectives.

BACKGROUND

Federal funding for highways is provided to the states mostly through a series of grant programs collectively known as the Federal-Aid Highway Program. Periodically, Congress enacts multiyear legislation that authorizes funding for the nation's surface transportation programs. In 2005, Congress enacted SAFETEA-LU, which authorized $197.5 billion for the Federal-Aid Highway Program for fiscal years 2005 through 2009. In a joint federal-state partnership, FHWA, within DOT, administers the Federal-Aid Highway Program and distributes most funds to the states through annual apportionments established by statutory formulas.[5] Once FHWA apportions these funds, the funds are available for obligation for construction, reconstruction, and improvement of highways and bridges on eligible federal-aid highway routes, as well as for other authorized purposes. The amount of federal funding made available for highways has been substantial—from $34.4 billion to $43 billion per year for fiscal years 2005 through 2009.

The Highway Trust Fund was established by Congress in 1956 to fund construction of the Interstate Highway System. The Highway Trust Fund receives excise taxes collected on motor fuels and truck-related taxes, including taxes on gasoline, diesel fuel, gasohol, and other fuels; truck tires and truck sales; and heavy vehicle use. The Department of the Treasury collects fuel taxes from a small number of corporations (i.e., oil refineries or fuel tank farms) located in a relatively small number of places and not directly from states. As such, FHWA calculates motor fuel-related contributions[6] based on estimates of the gallons of fuel used on highways in each state by relying on data gathered from state revenue agencies and summary tax data available from Treasury as part of the estimation process (see app. II).

Most of the funds from the Highway Account of the Highway Trust Fund (about 83.3 percent) were apportioned to states across 13 formula programs during the 5-year SAFETEA-LU period.[7] Included among these are 6 "core" highway programs (see table 1).

In addition to formula programs, Congress also directly allocated about 7.4 percent of Highway Account funds to state departments of transportation through congressionally directed High Priority Projects, which provided funding for a total of 5,091 specific projects identified in SAFETEA-LU. The remaining funds, about 9.3 percent of the total, represent dozens of other authorized programs allocated to state departments of transportation, congressionally directed projects other than High Priority Projects, administrative expenses, and funding provided to states by other DOT agencies, such as the National Highway Traffic Safety Administration and Federal Motor Carrier Safety Administration (see figure 1).

Table 1. Core Highway Programs

Program	Description
Congestion Mitigation and Air Quality	Funds the federal share of projects and programs to reduce transportation emissions in areas with poor air quality
Highway Bridge Program	Funds the federal share of projects to improve the condition of highway bridges through replacement, rehabilitation, and systematic preventive maintenance
Highway Safety and Improvement Program	Funds the federal share of projects designed to achieve a significant reduction in highway fatalities and serious injuries on public roads
Interstate Maintenance	Funds the federal share of projects to resurface, restore, rehabilitate, and reconstruct Interstate routes
National Highway System	Funds the federal share of projects improving roads that are part of the National Highway System
Surface Transportation Program	Funds the federal share of projects states and localities may carry out on any federal-aid highway, including bridge projects, transit capital projects and bus facilities

Source: FHWA.

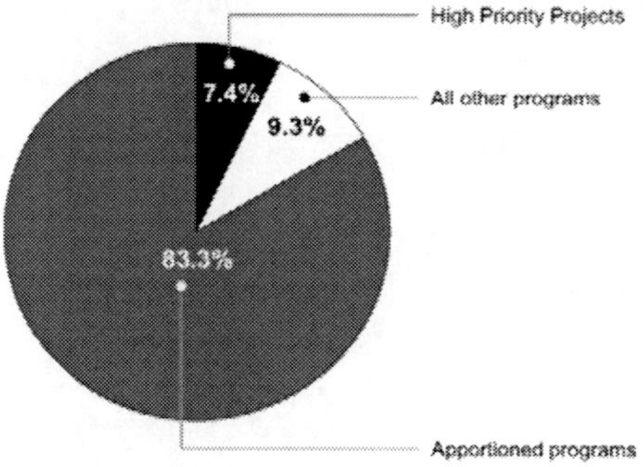

Source: GAO analysis of FHWA data.

Note: Apportioned programs funded through the Highway Account include: Interstate Maintenance, National Highway System, Surface Transportation, Congestion Mitigation and Air Quality, Highway Bridge, Appalachian Development Highway System, Coordinated Border Infrastructure, Highway Safety Improvement, metropolitan planning, Rail-Highway Safety, Recreational Trails, Safe Routes to Schools, and the Equity Bonus programs.

Figure 1. Highway Account Funds—Apportioned Programs, High Priority Projects, and All Other Programs, Fiscal Years 2005-2009.

Some of the apportioned programs use states' contributions to the Highway Account of the Highway Trust Fund as a factor in determining funding levels for each state.[8] As previously mentioned, FHWA has to estimate the fuel tax contributions made to the fund by users in each state. The collection and estimation process takes place over several years (see figure 2), and thus, the data used to calculate the formula are two years old. For example, the data used to apportion funding to states in fiscal year 2009 were based on estimated collections attributable to each state in fiscal year 2007.

By the early 1980s, construction of the Interstate Highway System was nearing completion, and a larger portion of the funds authorized from the Highway Trust Fund were for non-Interstate programs. The Surface Transportation Assistance Act of 1982 provided, for the first time, that each state would for certain programs receive a "minimum allocation" of 85 percent of its share of estimated tax payments to the Highway Account of the Highway Trust Fund. This approach was largely retained when Congress reauthorized the program in 1987. Since then, each state has received a specific share of the total program (defined as all apportioned programs plus High Priority Projects) and rate-of-return considerations into funds states received for the Interstate Maintenance, National Highway System, and Surface Transportation programs. In 2005, Congress implemented through SAFETEA-LU the Equity Bonus Program that was designed to give states a guaranteed rate of return of 92 percent by fiscal year 2008.

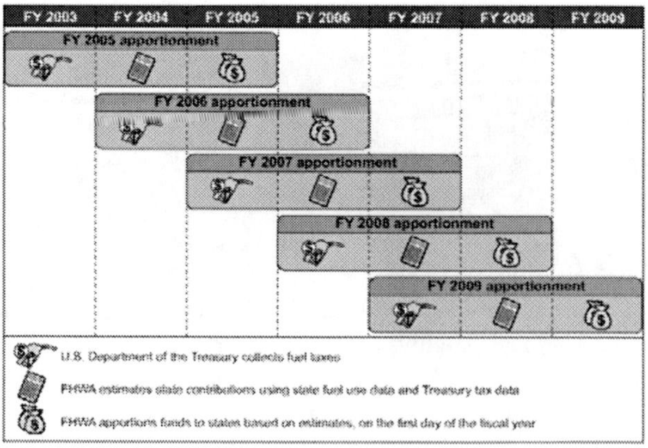

Source: GAO.

Figure 2. Time Lag between When Treasury Collects Fuel Taxes and Funds Are Apportioned.

In June 2010, we reported that every state but one received more funding for highway programs than users contributed to the Highway Account. The only exception, Texas, received about $1.00 (99.7 cents) for each dollar contributed. Among other states, this ranged from a low of $1.02 for both Arizona and Indiana to a high of $5.63 for the District of Columbia. In addition, all states, including Texas, received more funding than was contributed during both fiscal years 2007 and 2008. In effect, almost every state was a donee state during the first 4 years of SAFETEA-LU. This occurred because overall, more funding was authorized and apportioned than was collected from highway users, as the account was supplemented by general funds from Treasury.

We also reported that whether a state is viewed as a "donor" or "donee" state can depend on which of the four following methods of calculating states' rate of return is used:

1. **States' rate of return per dollar contributed to the Highway Account, using same-year comparison data.** This analysis compares funding states received from the Highway Trust Fund Highway Account with the dollars estimated to have been collected in each state into the Highway Account in that same year. For example, for fiscal year 2007, it compares the highway funds states received in fiscal year 2007 with the amount collected and contributed in that fiscal year. This analysis includes all funding provided to the states from the Highway Account, including (1) funds apportioned by formula, (2) High Priority Projects, and (3) other authorized programs, including safety program funding provided to states by other DOT agencies, such as the National Highway Traffic Safety Administration and Federal Motor Carrier Safety Administration (see figure 1 for a breakdown of these funds). This was the method used to calculate the results reported in our previous report that every state but one received more funding for highway programs than users contributed.
2. **States' rate of return per dollar contributed to the Highway Account, using time-lagged data, apportioned programs, and High Priority Projects.** This method applies the same dollar return calculation method but uses contribution data that were available at the time funds were apportioned to the states—the lagged-time data were 2 years old, due to the time lag between when Treasury collects fuel and truck excise taxes and funds are apportioned. It also uses a

subset of Federal-Aid Highway programs including both programs apportioned to states by formula and High Priority Projects; it does not include other allocated highway programs or other funding states receive from other DOT agencies, such as the National Highway Traffic Safety Administration and Federal Motor Carrier Safety Administration.

3. **States' relative share rate of return from the Highway Account, using time-lagged comparison data, apportioned programs, and High Priority Projects.** This third calculation method is based on a state's "relative share"—that is, the amount a state receives relative to other states instead of an absolute, dollar rate of return. In order to calculate this rate of return, FHWA must determine what proportion of the total national contributions came from highway users in each state. Each state's share of contributions into the Highway Account of the Highway Trust Fund is then used to calculate a relative rate of return—how the proportion of each state's contribution compares to the proportion of funds the state received.
4. **States' relative share rate of return per dollar contributed to the Highway Account, using same-year comparison data.** This fourth method for calculating a state's rate of return involves evaluating the relative share as described in method 3, but using the same-year comparison data.

States Received More Funding from the Highway Trust Fund Than Was Contributed from Fiscal Years 2005 through 2009, but Rate of Return Varies Depending on the Calculation Used

States Received More Funding Than Was Contributed from Fiscal Years 2005 through 2009

Our analysis of the entire 5-year period of SAFETEA-LU shows that every state was a donee state, receiving more funding for highway programs than their users contributed to the Highway Account (see figure 3). Funding received for each dollar contributed ranged from about $1.03 for Texas to about $5.85 for the District of Columbia. Every state was a donee state during the 5-year SAFETEA-LU period because overall, more funding was authorized and apportioned than was collected from highway users, since the account was supplemented by general funds from Treasury.[9] Since our June 2010 report, nearly every state increased its rate of return by at least 1 cent over what we originally reported. Hawaii was the only state that remained constant and did not have an increase on the rate of return in fiscal year 2009.

States' Rate of Return Varies, as other Methods provide different results

Although our analysis shows that states received more than was contributed, other calculations provide different results. Because there are different methods of calculating a rate of return, and the method used affects the results, confusion can result over whether a state is a donor or donee. A state can appear to be a donor using one type of calculation and a donee using a different type.

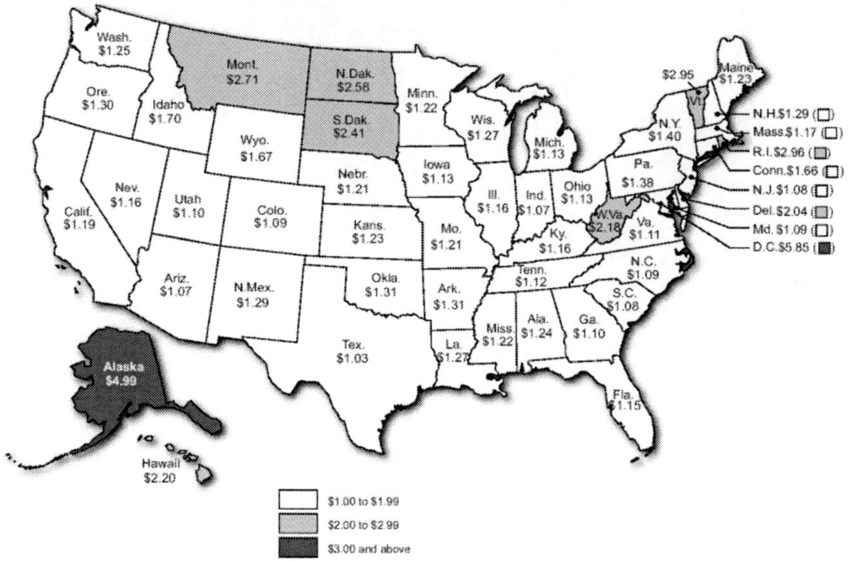

Sources: GAO analysis of FHWA data; Map Resources (map).

Figure 3. States' Rate of Return per Dollar Contributed to the Highway Account of the Highway Trust Fund, Using Same-Year Comparison Data, Fiscal Years 2005-2009.

We found that each state received more in funding than its users contributed during the SAFETEA-LU period when using time-lagged comparison data to calculate states' rate of return per dollar.[10] The rate of return ranged from a low of $1.04 per dollar for 16 states to a high of $5.26 per dollar for the District of Columbia, as shown in figure 4. This methodology results in states generally having a lower dollar rate of return than our calculations using same-year data and differs in that we use a subset of Federal-Aid Highway programs including both programs apportioned to states by formula and High Priority Projects.[11] The results from our June 2010 report have not changed because fiscal year 2009 data were included in our 2010 analysis, and all states are donee states.[12]

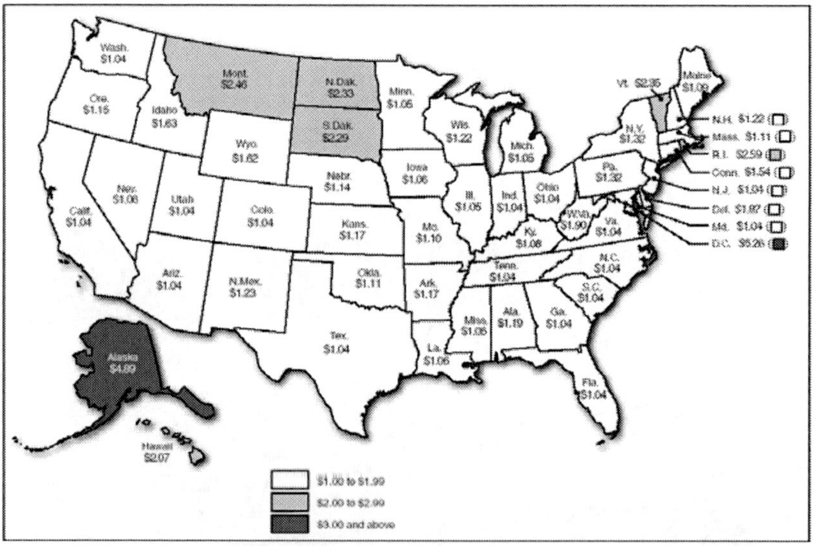

Sources: GAO analysis of FHWA data; Map Resources (map).

Note: Calculations compare the amount each state received through the apportioned programs and High Priority Projects in fiscal years 2005-2009 with each state's estimated contribution to the Highway Account in the corresponding revenue years (fiscal years 2003-2007).

Figure 4. States' Return per Dollar Contributed to the Highway Account of the Highway Trust Fund, Using Time-Lagged Data, Apportioned Programs, and High Priority Projects, Fiscal Years 2005-2009.

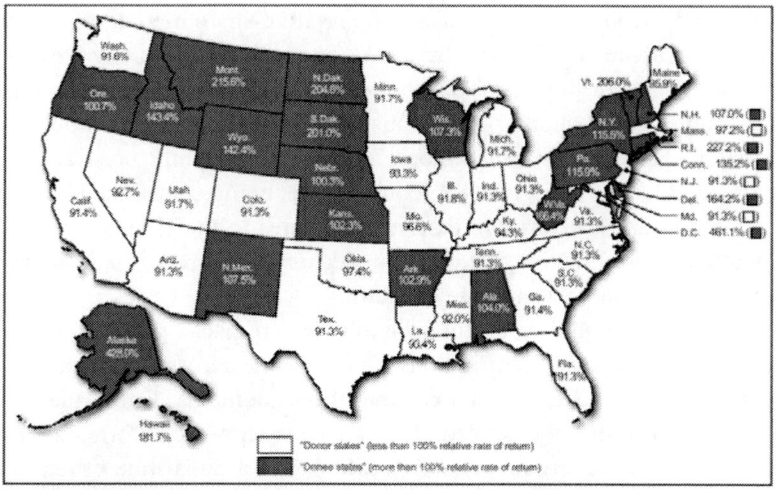

Sources: GAO analysis of FHWA data; Map Resources (map).

Note: Calculations compare the share of the total funding each state received through the apportioned programs and High Priority Projects in fiscal years 2005-2009 with the estimated share of the national total that each state contributed into the Highway Account in the corresponding revenue years (fiscal years 2003-2007).

Figure 5. States' Relative Share Rate of Return from the Highway Account of the Highway Trust Fund, Using Time-Lagged Comparison Data, Apportioned Programs, and High Priority Projects, Fiscal Years 2005-2009.

A third calculation, based on a state's "relative share"—the amount a state receives relative to other states using time-lagged data—results in both donor and donee states. Congress defined this method in SAFETEA-LU as the one FHWA uses for calculating rates of return for the purpose of apportioning highway funding to the states.[13] A comparison of the relative rate of return on states' contributions showed 28 donor states, receiving less than 100 percent relative rate of return, and 23 states as donees receiving more than a 100 percent relative rate of return (see figure 4). States' relative rates of return ranged from a low of 91.3 percent for 12 states to a high of 461 percent for the District of Columbia. Similar to the return per dollar analysis in figure 5, this calculation includes only apportioned funds and High Priority Projects allocated to states, and excludes other DOT authorized programs allocated to states (see figure 1). The difference between a state's absolute per dollar return and relative rate of return can create confusion because the relative share calculation is sometimes mistakenly referred to as "cents on the dollar."

As we previously reported, using the relative share method of calculation will result in some states being "winners" and other states being "losers."[14] If one state receives a higher proportion of highway funds than it is viewed as having contributed, another state must receive a lower proportion than it contributed. However, because more funding has recently been apportioned and allocated from the Highway Account than is being contributed by highway users, a state can receive more than it contributes to the Highway Trust Fund Highway Account, making it a donee under its rate of return per dollar, but a donor under its relative share rate of return.

California provides a useful example. From fiscal years 2005 through 2009, using same-year contributions and funding across all Highway Trust Fund Highway Account allocations and apportionments, California received $1.19 for each dollar contributed. This analysis shows California as a donee state (see table 2). Alternatively, when calculating a dollar rate of return using state contribution estimates available at the time of apportionment (as shown in figure 3) and including only programs covered in rate-of-return adjustments, California remains a donee state, receiving $1.04 for each dollar contributed. In contrast, using the relative share approach for the fiscal year 2005 through 2009 period, California received 91 percent of the share it contributed in federal highway-related taxes, which would make it a donor state.

Our fourth way of calculating a state's rate of return is not normally calculated by FHWA. It involves evaluating the relative share as just described, but using the same-year comparison data. Again, because of the time lag required to estimate state highway user contributions to the Highway Account, such analysis is possible only 2 years after FHWA calculates apportionments for states. Our analysis using this approach results in yet another set of rate-of-return answers. When this analysis is applied to all states, approximately half of the states are donor states and the other half are donee states. Specifically, under this methodology, 25 states received less than 100 percent relative rate of return, and 26 states received more than a 100 percent relative rate of return. Compared with the results we reported in our 2010 report, our recent analysis with fiscal year 2009 data resulted in Louisiana, Maine, and Washington changing from a donor to a donee state using the relative share method for a state's rate of return in the same year. For instance, in our June 2010 report, the relative share for Louisiana was 97.97 percent, but it increased to 102.98 percent with the inclusion of fiscal year 2009 data. Conversely, Minnesota shifted from a donor to a donee state. In our June 2010 report, the relative share for Minnesota was 101.26 percent but it changed to 99.46 percent with the inclusion of fiscal year 2009 data, again using the relative share method for a state's rate of return in the same year. Table 3 shows the results of all states for the entire 5-year SAFETEA-LU period using the four analysis methods.

Table 2. Comparison of Different Rates of Return for California from Highway Trust Fund Highway Account, by Method of Calculation

	Dollars	Share
Rate of return using same-year data		
California, fiscal years 2005-2009, received (includes all apportionments and allocations)	\$19.3 billion	
California, fiscal years 2005-2009, contributed	\$16.2 billion	
Calculation	\$19.3 billion ÷ \$16.2 billion	
Return per dollar using same-year comparison data, fiscal years 2005-2009	\$1.19	
Rate of return using data available at time of apportionment		
California, fiscal years 2005-2009, received (includes apportionments from 13 formula programs plus funds from the High Priority Project Program)	\$16.6 billion	9.07%
California, fiscal years 2003-2007, contributed	\$15.9 billion	9.93%
Calculation	\$16.6 billion ÷ \$15.9 billion	9.07% ÷ 9.93%
Rate of return using apportionment-year data, fiscal years 2005-2009	**\$1.04**	**91%**

Source: FHWA data.

Table 3. Comparison of States' Different Rates of Return from the Highway Trust Fund Highway Account, by Four Methods of Calculation

State	Return per dollar		Relative share	
	Same-year comparison, fiscal years 2005 through 2009	Year of apportionment comparison, fiscal years 2005 through 2009	Same-year comparison, fiscal years 2005 through 2009 (percentage)	Year of apportionment comparison, fiscal years 2005 through 2009 (percentage)
Alabama	\$1.24	\$1.19	101.05	103.98
Alaska	4.99	4.89	405.36	428.90
Arizona	1.07	1.04	86.98	91.30
Arkansas	1.31	1.17	106.42	102.89

Table 3. (Continued)

State	Return per dollar		Relative share	
	Same-year comparison, fiscal years 2005 through 2009	Year of apportionment comparison, fiscal years 2005 through 2009	Same-year comparison, fiscal years 2005 through 2009 (percentage)	Year of apportionment comparison, fiscal years 2005 through 2009 (percentage)
California	1.19	1.04	96.75	91.36
Colorado	1.09	1.04	88.67	91.31
Connecticut	1.66	1.54	135.24	135.24
Delaware	2.04	1.87	165.77	164.17
District of Columbia	5.85	5.26	475.64	461.14
Florida	1.15	1.04	93.53	91.32
Georgia	1.10	1.04	89.03	91.36
Hawaii	2.20	2.07	178.66	181.73
Idaho	1.70	1.63	137.86	143.38
Illinois	1.16	1.05	94.19	91.75
Indiana	1.07	1.04	86.89	91.30
Iowa	1.13	1.06	92.16	93.27
Kansas	1.23	1.17	100.17	102.31
Kentucky	1.16	1.08	93.90	94.34
Louisiana	1.27	1.06	102.98	93.42
Maine	1.23	1.09	100.10	95.90
Maryland	1.09	1.04	88.95	91.35
Massachusetts	1.17	1.11	95.35	97.22
Michigan	1.13	1.05	92.11	91.73
Minnesota	1.22	1.05	99.46	91.71
Mississippi	1.22	1.05	98.87	92.01
Missouri	1.21	1.10	98.12	96.60
Montana	2.71	2.46	219.95	215.56
Nebraska	1.21	1.14	97.95	100.31
Nevada	1.16	1.06	93.99	92.74
New Hampshire	1.29	1.22	104.64	106.98
New Jersey	1.08	1.04	87.81	91.34
New Mexico	1.29	1.23	104.92	107.49
New York	1.40	1.32	113.78	115.50
North Carolina	1.09	1.04	88.78	91.34
North Dakota	2.58	2.33	209.50	204.64
Ohio	1.13	1.04	92.00	91.33

State	Return per dollar		Relative share	
	Same-year comparison, fiscal years 2005 through 2009	Year of apportionment comparison, fiscal years 2005 through 2009	Same-year comparison, fiscal years 2005 through 2009 (percentage)	Year of apportionment comparison, fiscal years 2005 through 2009 (percentage)
Oklahoma	1.31	1.11	106.78	97.39
Oregon	1.30	1.15	105.88	100.72
Pennsylvania	1.38	1.32	112.37	115.91
Rhode Island	2.96	2.59	240.76	227.18
South Carolina	1.08	1.04	87.40	91.33
South Dakota	2.41	2.29	196.19	201.04
Tennessee	1.12	1.04	90.92	91.35
Texas	1.03	1.04	83.99	91.32
Utah	1.10	1.04	89.24	91.66
Vermont	2.95	2.35	239.44	206.04
Virginia	1.11	1.04	89.98	91.33
Washington	1.25	1.04	101.51	91.62
West Virginia	2.18	1.90	176.73	166.36
Wisconsin	1.27	1.22	103.11	107.27
Wyoming	1.67	1.62	135.41	142.35
Total	**$1.23**	**$1.14**	**100.00**	**100.00**

Source: GAO analysis of FHWA data

Note: Gray shading indicates donor state.

EQUITY BONUS PROVISIONS IN SAFETEA-LU ADDRESSED RATE OF RETURN AMONG STATES

The Equity Bonus provisions addressed states' concern on the rate of return and provided more funds to states than any other individual Federal-Aid Highway formula program. Over SAFETEA-LU's 5-year authorization period, the Equity Bonus provisions provided $44 billion in funding authority to the states, while the second-largest formula program, the Surface Transportation Program, provided $32.5 billion.

Each year since fiscal year 2005, about $2.6 billion stayed as Equity Bonus program funds and could be used for any purpose eligible under the Surface Transportation Program.

Any additional Equity Bonus funds are added to the apportionments of the six "core" Federal-Aid Highway formula programs: Interstate Maintenance, National Highway System, Surface Transportation, Congestion Mitigation and Air Quality, Highway Bridge, and Highway Safety Improvement. Because states can transfer funds among the core programs, the funding apportioned to any core program is not critical.

States may qualify for Equity Bonus funding by meeting criteria contained in one of three provisions (see figure 6). A state that meets the criteria for more than one of the provisions receives funding under the provision providing the greatest amount of funding. FHWA conducts Equity Bonus calculations annually.

However, with the extension of SAFETEA-LU authorization for 2 years, the Equity Bonus Program has not been recalculated. According to FHWA officials, since fiscal years 2010 and 2011 were funded at the fiscal year 2009 level, states simply received the same amount for fiscal years 2010 and 2011 as they did in fiscal year 2009.

For the first criterion—the guaranteed relative rate of return—for fiscal year 2005, all states were guaranteed at least 90.5 percent of their share of estimated contributions. The guaranteed rate of return increased in steps, rising to 92 percent in fiscal year 2009. The second criterion—the guaranteed increase over average annual apportionments authorized by the Transportation Equity Act for the 21st Century (TEA-21)[15]—also varied by year, rising from 117 percent in fiscal year 2005 to 121 percent in fiscal year 2009.[16] The number of states qualifying under the first two provisions varied from year to year. For the third criterion, a guarantee to "hold harmless" states that had certain qualifying characteristics at the time SAFETEA-LU was enacted, 27 states had at least one of these characteristics. A number of states had more than one of these characteristics.

Forty-seven states received Equity Bonus funding every year during the SAFETEA-LU period. However, the District of Columbia, Rhode Island, and Vermont each had at least 1 year where they did not receive Equity Bonus funding because they did not need it to reach the funding level specified under the three provisions. Maine was the only state that did not receive an Equity Bonus in any year. As a result, half of all states received at least a 25 percent increase in their overall Federal-Aid Highway Program funding over their core funding. Each state's percentage increase in its overall funding total for apportioned programs and High Priority Projects for fiscal years 2005 through 2009 due to Equity Bonus funding is shown in figure 7.

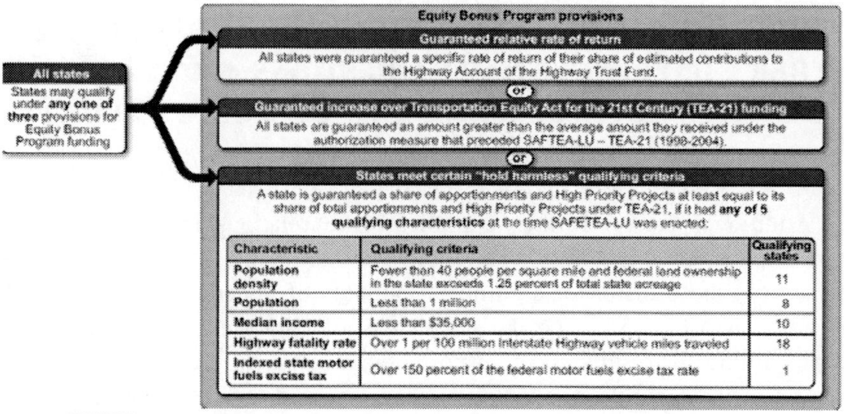

Source: FHWA.

Note: Because states may have met more than one qualifying criteria under the hold harmless provision, the qualifying state column is not additive.

Figure 6. Equity Bonus Program Criteria.

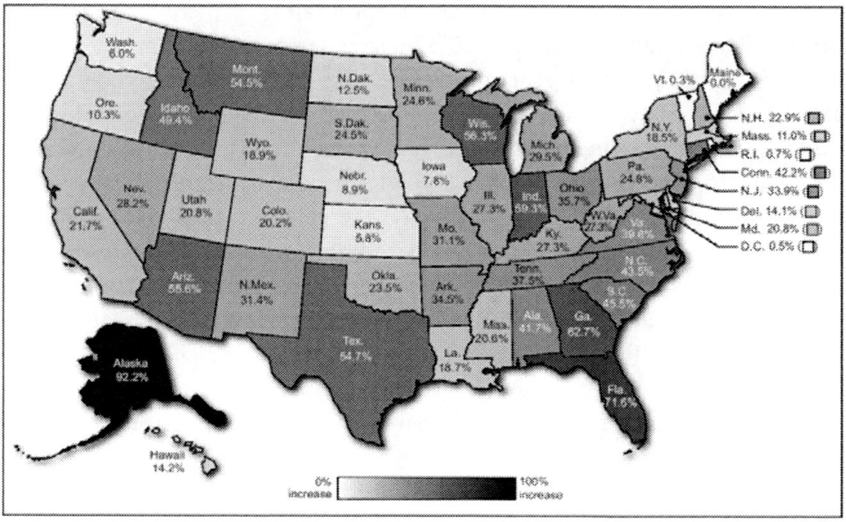

Sources: GAO analysis of FHWA data; Map Resources (map).

Figure 7. Percentage Increase in Total State Apportionment and High Priority Amounts Due to Equity Bonus, Fiscal Years 2005-2009.

ADDING GENERAL REVENUES INTO THE TRUST FUND AND OTHER CHALLENGES RAISE QUESTIONS ABOUT RELYING ON STATES' RATE OF RETURN TO DISTRIBUTE FEDERAL HIGHWAY FUNDS

Additional factors further complicate the relationship between states' contributions to the Highway Trust Fund and the funding states receive. These include (1) the infusion of significant amounts of general revenues into the Highway Trust Fund, (2) the challenge of factoring performance and accountability for results into transportation investment decisions, and (3) the long-term sustainability of existing mechanisms and the challenges associated with developing new approaches to funding the nation's transportation system.

The infusion of significant amounts of general revenues into the Highway Trust Fund Highway Account breaks the link between highway taxes and highway funding. In fiscal year 2008, the Highway Trust Fund held insufficient amounts to sustain the authorized level of funding and, partly as a result, we placed it on our list of high-risk programs.[17] To cover the shortfall, from fiscal years 2008 through 2010 Congress approved the transfer of about $29.7 billion in additional general revenues into the Highway Account of the Highway Trust Fund.[18] This transfer affected each state's rate of return and resulted in all states being donees—receiving more funds than they contributed to the Highway Account. Taken as a whole, for fiscal year 2009, the general fund transfers had a significant impact on the states' rate of return, with the federal government paying about $42.4 billion to the states, while highway user fees paid into the Highway Account were $30.1 billion. This means that, to a large extent, funding has shifted away from the contributions of highway users, breaking the link between highway taxes paid and benefits received by users.[19] Furthermore, the infusion of a significant amount of general fund revenues complicates rate-of-return analysis because the current method of calculating contributions does not account for states' general revenue contributions. Because for many states the share of Highway Trust Fund contributions and general revenue contributions are different, state-based contributions to all the funding in the Trust Fund are no longer clear.[20] In addition, since March 2009, the American Recovery and Reinvestment Act of 2009 apportioned an additional $26.7 billion to the states for highways—a significant augmentation of federal highway spending that was funded with general revenues.[21]

Using rate of return as a major factor in determining federal highway funding levels is at odds with re-examining and restructuring federal surface transportation programs so that performance and accountability for results is factored into transportation investment decisions. As we have reported, for many surface transportation programs, goals are numerous and conflicting, and the federal role in achieving the goals is not clear. Many of these programs have no relationship to the performance of either the transportation system or the grantees receiving federal funds and do not use the best tools and approaches to ensure effective investment decisions.[22] Our previous work has outlined the need to create well-defined goals based on identified areas of federal interest and a clearly defined federal role in relation to other levels of government.[23] We have suggested that where the federal interest is less evident, state and local governments could assume more responsibility, and some functions could potentially be assumed by the states or other levels of government.[24] Furthermore, incorporating performance and accountability for results into transportation funding decisions is critical to improving results. However, the current approach presents challenges. The Federal-Aid Highway Program, in particular, distributes funding through a complicated process in which the underlying data and factors are ultimately not meaningful because they are overridden by other provisions designed to yield a largely predetermined outcome—that of returning revenues to their attributed state of origin.[25] Moreover, once the funds are apportioned, states have considerable flexibility to reallocate them among highway and transit programs.[26] As we have reported, this flexibility, coupled with a rate-of-return orientation, essentially means that the Federal-Aid Highway Program functions, to some extent, as a cash transfer, general purpose grant program.[27] This approach poses considerable challenges to introducing performance orientation and accountability for results into highway investment decisions. For three highway programs that were designed to meet national and regional transportation priorities, we have recommended that Congress consider a competitive, criteria-based process for distributing federal funds.[28]

Finally, using rate of return as a major factor in determining federal highway funding levels poses problems because funding the nation's transportation system through taxes on motor vehicle fuels is likely unsustainable in the long term. Receipts for the Highway Trust Fund derived from motor fuel taxes have declined in purchasing power, in part because the federal gasoline tax rate has not increased since 1993. In fiscal year 2008, total contributions to the Highway Account of the Highway Trust Fund decreased by more than $3.5 billion from fiscal year 2007, the first year of decrease during the SAFETEA-LU period. The Congressional Budget Office forecasts another revenue shortfall in the Highway Account of the Highway Trust Fund by the end of fiscal year 2012.[29] Over the long term, vehicles will become more fuel efficient and increasingly run on alternative fuels. As such, fuel taxes may not be a long-term source of transportation funding.[30] Furthermore, transportation experts have noted that transportation policy needs to recognize emerging national and global challenges, such as reducing the nation's dependence on imported fuel and minimizing the effect of transportation systems on the global climate.[31] A fund that relies on increasing the use of motor fuels to remain solvent might not be compatible with the strategies that may be required to address these challenges.

In the future, policy discussions will need to consider what the most adequate and appropriate transportation financing systems will be, and whether the current system continues to make sense. The National Surface Transportation Infrastructure Financing Commission—created by SAFETEA-LU to, among other things, explore alternative funding mechanisms for surface transportation—identified and evaluated numerous revenue sources for surface transportation programs in its February 2009 report, including alternative approaches to the fuel tax, mileage-based user fees, and freight-related charges.[32] The report also discussed using general revenues to finance transportation investment but concluded that it was a weak option in terms of economic efficiency and other factors and recommended that new sources of revenue to support transportation be explored. These new sources of revenue may or may not lend themselves to using a rate-of-return approach.

Agency Comments and Our Evaluation

We provided a draft of this report to DOT for review and comment. DOT provided technical comments, which we incorporated as appropriate.

Phillip R. Herr

Director, Physical Infrastructure Issues

Appendix I. Objectives, Scope and Methodology

This report addresses the following objectives: (1) the amount of revenue contributed to the Highway Trust Fund Highway Account compared with the funding states received during the Safe, Accountable, Flexible, Efficient Transportation Equity Act: A Legacy for Users (SAFETEA-LU) period; (2) the provisions in place during the SAFETEA-LU period to address rate-of-return issues across states, and how they affected the highway funding states received; and (3) additional factors that affected the relationship between contributions to the Highway Trust Fund and the funding states receive.

This report updates and includes additional information for a related report issued in June 2010.[33] The main update to the prior report is the inclusion and additional analysis of fiscal year 2009 data on the states' contributions to the Highway Trust Fund that was made publicly available by Federal Highway Administration (FHWA). As a result, the analyses in this update are inclusive of the entire initial 5-year SAFETEA-LU period. Any analyses from the 2010 report that were not impacted by new data were included in this update as context.

To determine the amount of revenue states contributed to the Highway Trust Fund Highway Account compared with the funding they received during the SAFETEA-LU period, we completed four analyses using FHWA data. We met with FHWA and other Department of Transportation (DOT) officials to discuss availability of data and appropriate methodologies. We used FHWA estimates of payments made into the Highway Account of the Highway Trust Fund, by state, and the actual total apportionments and allocations made from the fund, by state. This is sometimes referred to as a "dollar-in, dollar-out" analysis. The source data is published annually in Highway Statistics as Table FE-221, titled "Comparison of Federal Highway Trust Fund Highway Account Receipts Attributable to the States and Federal-Aid Apportionments and Allocations from the Highway Account." FHWA officials confirmed that it contains the best estimate of state contributions and actual total appropriations and allocations received by states from the Highway Account of the fund. We did not independently review FHWA's process for estimating states' contributions into the Highway Trust Fund. However, we have reviewed this process in the past, and FHWA made changes to the process as a result of our review.[34] In addition, we did not attribute any prior balances in the Highway Trust Fund back to states of origin because these funds are not directly tied to any specific year or state.

We performed alternative analyses to demonstrate that different methodologies provide different answers to the question of how the contributions of states' highway users compared with the funding states received. Using the same data as just described, we performed a "relative share" analysis, which compared each state's estimated proportion of the total contributions to the Highway Account with each state's proportion of total Federal-Aid Highway funding. We also examined how states fared using FHWA's approach, for determining the Equity Bonus Program funding apportionments. We performed this analysis to show the outcomes for states based on the information available at the time the Equity Bonus program apportionments are made. The Equity Bonus program amounts are calculated using the statutory formulas for a subset of Federal-Aid Highway Programs. These include all programs apportioned by formula plus the allocated High Priority Projects. FHWA uses the most current contribution data available at the time it does its estimates. However, as explained, the time lag for developing this data is about 2 years. Therefore, we applied the contribution data for 2003 through 2007 to the funding data for 2005 through 2009, the full SAFETEA-LU period. For these data, we analyzed (1) the total estimated contributions by state divided by the total funding received by state—the dollar-in, dollar-out methodology—and (2) a comparison of the share of contributions to share of payments received for each state. We obtained data from the FHWA Office of Budget for the analysis of state dollar-in, dollar-out outcomes and state relative share data for the Equity Bonus Program. We completed our analyses across the total years of the SAFETEA-LU period, 2005 through 2009. We interviewed FHWA officials and obtained additional information from FHWA on the steps taken to ensure data reliability and determined the data were sufficiently reliable for the purposes of this report.

To determine the provisions in place during the SAFETEA-LU period to address rate-of-return issues across states, and how they affected the highway funding states received, we reviewed SAFETEA-LU legislation and reports by the Congressional Research Service (CRS), FHWA, and others as applicable. We conducted an analysis of the of FHWA data on the Equity Bonus Program provisions, which were created explicitly to address the rate-of-return issues across states. Our analysis compared funding levels distributed to states through apportionment programs and High Priority Projects before and after Equity Bonus Program provisions were applied, and calculated the percentage increase each state received as a result of the Equity Bonus. We also spoke with FHWA officials to get their perspectives.

To determine what additional factors affected the relationship between contributions to the Highway Trust Fund compared with the funding states receive, we reviewed GAO reports on the Highway Trust Fund and federal surface transportation programs, CRS and FHWA reports, and the National Surface Transportation Infrastructure Financing Commission report. In addition, we reviewed existing FHWA data on the status of the Highway

Account of the Highway Trust Fund. We also met with officials from DOT's Office of Budget and Programs and FHWA to obtain their perspectives on the issue.

APPENDIX II. HOW FHWA ESTIMATES CONTRIBUTIONS TO THE HIGHWAY TRUST FUND

Currently, the Federal Highway Administration (FHWA) estimates state-based contributions to the Highway Account of the Highway Trust Fund through a process that includes data collection, adjustment, verification, and final calculation of the states' highway users' contributions. First, FHWA collects monthly motor fuel use data and related annual state tax data from state departments of revenue. FHWA then adjusts states' data by applying its own models using federal and other data to establish data consistency among the states. FHWA provides feedback to the states on these adjustments and estimates through FHWA Division Offices. Finally, FHWA applies each state's estimated share of highway fuel usage to total taxes collected nationally to arrive at a state's contribution to the Highway Trust Fund. We did not assess the effectiveness of FHWA's process for estimating the amount of tax funds attributed to each state for this report.[35]

According to FHWA officials, because data from state revenue agencies are more reliable and comprehensive than vehicle miles traveled data, FHWA uses state tax information to calculate state contributions. States submit regular reports to FHWA, including a monthly report on motor fuel consumption due 90 days after month's end and an annual report on motor fuel tax receipts due 90 days after calendar year's end. Because states have a wide variety of fuel tracking and reporting methods, FHWA must adjust the data to achieve uniformity. FHWA analyzes and adjusts fuel usage data, such as off-highway use related to agriculture, construction, industrial, marine, rail, aviation, and off-road recreational usage. It also analyzes and adjusts use data based on public sector use, including federal civilian, state, county, and municipal use.

FHWA headquarters and division offices also work together to communicate with state departments of revenue during the attribution estimation process.[36] According to FHWA officials, each year FHWA headquarters issues a memo prompting its division offices to have each state conduct a final review of the motor fuel gallons reported by their respective states. FHWA division offices also are required to assess their state's motor fuel use and highway tax receipt process at least once every 3 years to determine if states are complying with FHWA guidance on motor fuel data collection.

Once the data are finalized, FHWA applies each state's estimated share of taxed highway fuel use to the total taxes collected to arrive at a state's contribution in the following manner. Finalized estimations of gallons of fuel used on highways in two categories—gasoline and special fuels—allow FHWA to calculate each state's share of the total on-highway fuel usage. The shares of fuel use for each state are applied to the total amount of taxes collected by the Department of the Treasury in each of the 10 categories of highway excise tax. The state's gasoline share is applied to the gasoline and gasohol taxes, and the state's special fuels share, which includes diesel fuel, is applied to all other taxes, including truck taxes.[37]

End Notes

[1] Pub. L. No.109-59, § 1101, 119 Stat. 1144, 1153 (2005).

[2] GAO, *Highway Trust Fund: Nearly All States Received More Funding Than They Contributed in Highway Taxes Since 2005*, GAO-10-780 (Washington, D.C.: June 30, 2010).

[3] Funding authorized by SAFETEA-LU was for fiscal year 2005 through fiscal year 2009. Congress has extended the authorization, and the current extension is set to expire on September 30, 2011.

[4] GAO-10-780.

[5] Our discussion of states in this report includes the District of Columbia.

[6] While we refer to these contributions as state Highway Trust Fund contributions, these are not based on actual contributions by the states, or taxes collected by them, or taxes paid by their residents, but rather an estimated allocation of national highway tax revenues based on the analytical model used by FHWA in calculating state bonuses for purposes of 23 U.S.C. § 105 (see app. II).

[7] The Highway Trust Fund is divided into the Highway Account and the Mass Transit Account. 26 U.S.C § 9503. More than 80 percent of the total revenue in the Highway Trust Fund is dedicated to the Highway Account.

[8] These programs include Interstate Maintenance, Surface Transportation, and Equity Bonus.

[9] As states received more in funding than highway users contributed in taxes, both an existing balance in the Highway Account was drawn down, and the account was supplemented by other funds from Treasury. We did not attempt to estimate what residual contribution in these prior balances is theoretically attributable to a state in our analysis.

[10] As previously mentioned, the method of using time-lagged data consists of calculating funding states received from the Highway Trust Fund Highway Account with estimated dollars that states contributed at the time funds are apportioned to the states. States' contribution data that are available at the time funds were apportioned to the states are 2-year-old lagged-time data, due to the time lag between when Treasury collects the fuel and truck excise taxes and funds are apportioned (see figure 2).

[11] This set of programs is used in the calculation of Equity Bonus Program funding, discussed later in this report.

[12] As previously noted, the second method of calculating states' rate of return consists of using 2-year-old data on contributions for apportionments, due to the time lag between when Treasury collects fuel and truck excise taxes and funds are apportioned. Apportionments are made on the first day of the fiscal year, and at that time, tax collections for the year are not known. Likewise, tax collections for the year that ended the day before the apportionments are not known.

[13] These FHWA calculations are part of the Equity Bonus Program, discussed later in the report.

[14] GAO-10-780.

[15] Pub. L. No. 105-178, title I, § 1104, 112 Stat 107, 127-129 (1998).

[16] 23 U.S.C. § 105(c).

[17] GAO, *High Risk Series: An Update*, GAO-09-271 (Washington, D.C.: January 2009).

[18] The $29.7 billion transferred amount was about 35 percent of FHWA's budget for fiscal years 2009 and 2010 combined. Robert S. Kirk, Congressional Research Service, *The Donor-Donee State Issue in Highway Finance*, (Washington, D.C.: June 13, 2011).

[19] The rate-of-return approach was designed to ensure that, consistent with the user pay system—wherein the costs of building and maintaining the system are borne by those who benefit—users receive a fair return on their investment to the extent possible.

[20] This also complicates longer-term historical analyses of state-based rate of return for the Highway Account. During certain periods, general funds from Treasury have been added to the Highway Trust Fund in the form of interest payments on the Highway Trust Fund balance. Conversely, in the past, fuel taxes have also been used for deficit reduction.

[21] Rate of return does not apply to American Recovery and Reinvestment Act funding and was not used in the rate of return calculations for this report. GAO, *Recovery Act: Funding Used for Transportation Infrastructure Projects, but Some Requirements Proved Challenging*, GAO-11-600 (Washington, D.C.: June 29, 2011).

[22] GAO, *Surface Transportation: Restructured Federal Approach Needed for More Focused, Performance-Based, and Sustainable Programs*, GAO-08-400 (Washington, D.C.: Mar. 6, 2008).

[23] GAO, *Surface Transportation Programs: Proposals Highlight Key Issues and Challenges in Restructuring the Programs*, GAO-08-843R (Washington, D.C.: July 29, 2008).

[24] GAO-08-400.

[25] GAO, *Federal-Aid Highways: Trends, Effect on State Spending, and Options for Future Program Design*, GAO-04-802 (Washington, D.C.: Aug. 31, 2004).

[26] GAO, *Surface Transportation: Principles Can Guide Efforts to Restructure and Fund Federal Program*, GAO-08-744T (Washington, D.C.: July, 10, 2008).

[27] GAO-04-802.

[28] GAO, *Surface Transportation: Clear Federal Role and Criteria-Based Selection Process Could Improve Three National and Regional Infrastructure Programs*, GAO-09-219 (Washington, D.C.: Feb. 6, 2009).

[29] Congressional Budget Office, *The Highway Trust Fund and Paying for Highways* (Washington, D.C.: May 17, 2011).

[30] GAO, *Highway Trust Fund: Improved Solvency Mechanisms and Communication Needed to Help Avoid Shortfalls in the Highway Account*, GAO-09-316 (Washington, D.C.: Feb. 6, 2009).

[31] GAO-08-843R.

[32] National Surface Transportation Infrastructure Financing Commission, *Paying Our Way: A New Framework for Transportation Finance* (Feb. 26, 2009).

[33] GAO-10-780.

[34] GAO, *Highway Funding: Problems with Highway Trust Fund Information Can Affect State Highway Funds*, GAO-RCED/AIMD-00-148 (Washington, D.C.: June 2000).

[35] We last reviewed this process in 2000. See GAO-RCED/AIMD-00-148. At that time, we made a number of recommendations to improve the process, and FHWA instituted improvements based on these recommendations.

[36] FHWA maintains a division office in each of the states and the District of Columbia.

[37] Special fuels principally includes diesel fuel but also includes very small amounts of highway uses of liquefied petroleum gas, kerosene, natural gas, and biodiesel.

INDEX

D

E

F

G

H

I

Q

R

S

T

U

V

W

Y